# A Traveller's Guide to GEOLOGICAL WONDERS *in Alberta*

Ron Mussieux

Marilyn Nelson

THE PROVINCIAL MUSEUM OF ALBERTA

Federation of Alberta Naturalists

**Canadian Cataloguing in Publication Data**

Mussieux, Ron, 1949-

A traveller's guide to geological wonders in Alberta

Copublished by: Federation of Alberta Naturalists, and Canadian Society of Petroleum Geologists.

Includes bibliographical references and index.

ISBN 0-7785-0123-X

1. Geology--Alberta--Guidebooks. 2. Alberta--Guidebooks. I. Nelson, Marilyn, 1955- II. Provincial Museum of Alberta. III. Federation of Alberta Naturalists. IV. Canadian Society of Petroleum Geologists. V. Title.

QE186.M87 1998 557.123 C98-910298-X



*Printed in Canada*

# CONTENTS

## CHAPTER 3 EDMONTON AND AREA



## CHAPTER 4 ALBERTA'S HEARTLAND



## CHAPTER 5 ALBERTA'S ROCKIES

# FOREWORD

Alberta is a land of contrasts. We are justifiably famous for our Rocky Mountains, but their grandeur comes, in part, from the contrast to our arid plains, productive parkland, and vast boreal forest. Travellers around Alberta would expect a geological guide to dwell extensively on the mountains but the evidence of geological forces is everywhere. Ron Mussieux and Marilyn Nelson have for the first time compiled stories of about 110 geological sites in this book. Sites are found in every ecoregion and tourist zone in the province. They are found in remote, distant locations or in our largest cities.

The challenge to writing any book about geological forces is to avoid overwhelming your readers with background information — the physics, chemistry, chronology and jargon of geology. Ron and Marilyn have modelled this book after successful guides used for bird watchers and plant enthusiasts. This is not a text book, it is a field guide, something to take with you on day trips or long vacations around Alberta. The photos, color drawings, superb maps and text combine to open the world of the geologist to everybody. Each site has its own fascinating story but these are linked to background information that paints a picture of the geological history of Alberta that is both compelling and easy to understand.

Ron has been Curator of Geology at the Provincial Museum of Alberta for 25 years. Marilyn has worked for the Museum on a variety of geological projects over the last 15 years. They have been in the business of communicating geology to the public for years. Museum displays cannot do justice to geological sites. This book allows Ron and Marilyn to open a new world to travellers that they never could to museum visitors. Their love of, and appreciation for, Alberta is apparent in the guide.

The Provincial Museum of Alberta is pleased to have worked with the Federation of Alberta Naturalists, the Canadian Society of Petroleum Geologists, and other partners on this project. Their involvement has been critical to the success of the guide. I am personally excited about the book as it fills a big gap in the literature on Alberta — we have wonderful guides to the plants and animals of the province and now we have one about the wonders of geology.

Do you know the best part about a book about geological sites? When you go to look at one of the sites covered in the book, unlike a bird or a mammal, you can count on it being there!

Enjoy your discovery of some of the real wonders of Alberta.

*Dr. W. Bruce McGillivray*
*Provincial Museum of Alberta*

## ACKNOWLEDGMENTS

We would like to thank the Provincial Museum of Alberta and the Department of Community Development for providing us the opportunity to work on this project. We are especially grateful to Dr. Bruce McGillivray for helping to develop the concept for the book and coordinating the project.

This book could not have been written without the assistance of many individuals, professional societies and institutions. We are grateful for funding provided by the Provincial Museum of Alberta, the Federation of Alberta Naturalists, the Canadian Society of Petroleum Geologists, the Canadian Geological Foundation, the Logan Fund of the Geological Association of Canada, and the Edmonton Geological Society. We would also like to express our gratitude to the Alberta Geological Survey (Alberta Energy and Utilities Board) for providing the expertise of Dan Magee, whose illustrations greatly enhance the interpretive and educational value of many of the site descriptions. The help and advice of Dixon Edwards, also of the Geological Survey, was instrumental in the development of the book.

Dr. Bruce McGillivray, Andrew Locock, Dr. Sam Nelson, Dr. Harold Bryant, Wayne Hill and Gil Freschauf all read and commented constructively on the book. Their reviews were greatly appreciated and helped to improve the product. A special thanks to Dr. Jim Burns for his tireless and painstaking editing of the final text.

Many people contributed ideas, photographs, and helped with initial research. Special thanks go to Alwynne Beaudoin, Jack Brink, Jim Burns, Julie Hrapko, Michael Llewelyn, Jane Ross (Provincial Museum of Alberta); Thomas Chacko, David Cruden, Brett Purdy, Rob Young (University of Alberta); Dixon Edwards, Richard Stein (Alberta Geological Survey); Adam Hedinger, John Kramers (Alberta Research Council), William Last (University of Manitoba), Grant Mossop (Geological Survey of Canada), Samuel J. Nelson (University of Calgary), Milt Wright (British Columbia Archaeology Branch), Banff National Park, Glenbow Library and Archives, Jasper National Park, Natural Resources Service (Alberta Environmental Protection), Provincial Archives of Alberta, Royal Tyrrell Museum of Palaeontology, Syncrude Canada Inc., and Wood Buffalo National Park.

The location maps were produced by Wendy Johnson of Johnson Cartographics Inc. We gratefully acknowledge the layout and design skills of Carolyn Lilgert of the Provincial Museum of Alberta.

Lastly, we add our thanks to the many geologists who did the original research for these sites. Our sources of information include hundreds of geological reports and papers, a number of specialist field guides, and personal contacts with a number of our colleagues.

# DISCLAIMER

*This book has been prepared by the Provincial Museum of Alberta as a public service to increase public appreciation of Alberta's geological wonders. Travelling to and exploring any of the geological sites described in this book has inherent risks, some of which are hidden. These risks include, but are not limited to, vehicular traffic, inclement weather, water, fire, cliffs, waterfalls, animals, plants, insects, slippery, unstable surfaces and poisonous substances that may cause serious injury or death. Proper planning, a working knowledge of the area and common sense are required before undertaking any outdoor expeditions.*

*The Provincial Museum of Alberta, Department of Community Development, Federation of Alberta Naturalists, Canadian Society of Petroleum Geologists and the authors disclaim any liability in negligence or otherwise for any loss, injury or damage which may occur as a result of reliance on any information contained in this book.*

# CHAPTER 1: INTRODUCTION

*Surface erosion of sandstone by running water, Dinosaur Provincial Park.*
*Ron Mussieux – Provincial Museum of Alberta*

# INTRODUCTION

With a great deal of encouragement from the Provincial Museum of Alberta, other government departments, and geological societies, we set out to write a book about Alberta's geology for the general public. Alberta has a surprising geological diversity and some of the most spectacular scenery anywhere.

This book is not a textbook but, instead, is a compendium of over one hundred described geological sites complete with maps to help you find them. Colorful, informative line diagrams produced by Dan Magee greatly enhance the interpretive text. To assist those who are more interested in the subject of geology, we have included a short primer to introduce some basic concepts.

We hope this book will encourage people to travel about the province and look at rocks and landforms with new insight. To assist travellers in planning their trips we have organized the book into the six tourism destination regions, each with its own chapter and a regional map showing all of our selected sites within the region.

Of course, Alberta has thousands of potential geological sites to choose from. Our selection of the sites was based on a number of factors including: personal knowledge, perceived geological importance, completeness, and general accessibility. We have included many sites within Alberta's Rocky Mountain national parks where there is abundant interpretive information, and we have also included many lesser known sites from the rest of the province. In addition, there are some sites included, particularly in the north, where access can be gained only by air or water. We felt these sites would still be of interest to the public.

We have chosen a variety of geological sites including significant landforms such as the Morley drumlins or the Athabasca sand dunes; outstanding rock outcrops such as the Crowsnest Volcanics or the Grassi Lakes reef; closed and operating quarries or industrial sites of geological significance such as the Leduc discovery well and the Syncrude Oil Sands plants; a few mineral collecting sites; and some important fossil sites.

It is our hope that our efforts will be rewarded by a better informed public with an interest in Alberta's geological history.

# PLANNING YOUR EXCURSION

## MAPS AND ACCOMMODATION

Once you have chosen the area of Alberta you plan to visit, check in the book to find which tourist destination region it lies in and which sites are along your route. You might want to obtain accommodation guides — the Alberta Motor Association and Travel Alberta outlets have up-to-date publications. If the site is in a remote area or requires hiking to reach it, we advise purchasing a detailed topographic map (1:250,000 or 1:50,000 scale) from Map Town. Even though location maps are provided for each site and were accurate at the time of printing, road numbers and roads do change over time. Make sure you get the most recent highway map possible.

## ACCESS

Only a few of the sites are somewhat difficult to reach. Such sites are included in the book because we felt they were of sufficient importance or of special interest. Although most of the sites are on government land, a few are on private property. Respect "No Trespassing" signs and be sure to ask land owners for permission to enter their property.

## SAFETY

Safety should be your first priority whether the site you visit is on the highway or at the end of a difficult trail. Accidents can happen anywhere but they are particularly unfortunate if they spoil a vacation. Visiting geological sites is an outdoor activity and there are dangers such as slippery rocks near waterfalls, steep and crumbly cliffs and slopes, crevasses in glaciers, highway traffic, water hazards, and unstable structures at abandoned mines sites. Most of the sites in this book have no interpretive signage alerting you to hazards, but they all have inherent risks.

Some sites, such as those in the mountains or Alberta's north, are in remote wilderness areas. Carry adequate survival clothing, food, and water. Be prepared for inclement weather and unexpected, and possibly dangerous, wildlife guests, such as bears.

Water safety is an important issue, especially in isolated whitewater localities. Boating on some of northern Alberta's more remote rivers should be attempted only by skilled canoeists with abundant wilderness experience. In rivers with rapids, canoes are the most practical watercraft because they can hold adequate supplies and they can be portaged. Water safety in remote areas takes special planning and we advise that you fill out a wilderness travel registration form with local RCMP, and inform a relative or friend about where you are headed. Finally, do not boat alone; follow water safety rules; and remember, no one has ever drowned taking a portage!

## COLLECTING

Collecting specimens is another important issue. Generally, there are no restrictions to the casual collecting of rocks and minerals, provided permission by the land owners has been obtained. However, crown reserved lands, such as Provincial Parks, National Parks, Ecological Reserves, and Wilderness Areas, do have their own regulations, and in most cases collecting is not allowed without written permission from the appropriate government authorities. In particular, the Alberta Historical Resources Act has regulations regarding fossil collecting. You can collect loose fossils lying on the ground, but you cannot excavate unless you have obtained a special permit. By law, all fossils collected are owned by the province. Collecting in active mine sites and quarries is prohibited. For the enjoyment of future visitors, please minimize the number of rock samples you remove from the area.

Enjoy your geological trek across Alberta. It is truly a unique province and deserves its reputation for its geological variety and beauty.

*Ron Mussieux*
*Marilyn Nelson*

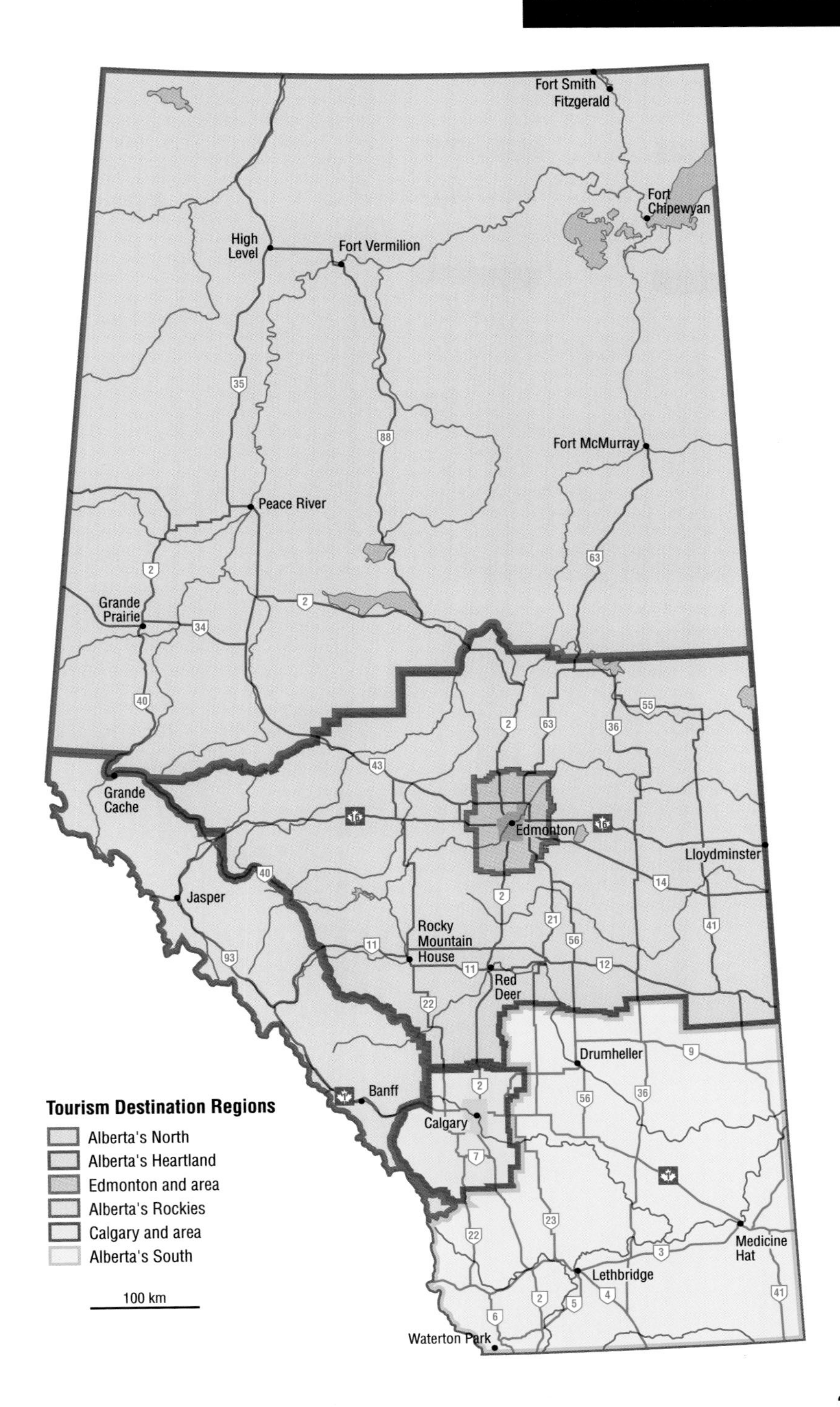
Fort Smith
Fitzgerald
Fort Chipewyan
High Level
Fort Vermilion
35
88
Fort McMurray
Peace River
63
2
Grande Prairie
34
40
55
36
43
Grande Cache
16
Edmonton
Lloydminster
14
Jasper
11
Rocky Mountain House
21
56
41
93
12
Red Deer
22
Drumheller
9
Banff
Calgary
7
23
Medicine Hat
3
Lethbridge
4
5
6
Waterton Park
Tourism Destination Regions
Alberta's North
Alberta's Heartland
Edmonton and area
Alberta's Rockies
Calgary and area
Alberta's South
100 km

# HOW TO USE THIS BOOK

Color bar indicates tourist destination region

## THE STRANGE SPHERES OF RED ROCK COULEE

Site name

*Natural Resources Service*

### HIGHLIGHTS

Concise, abbreviated highlights of the site

The eerie moon-like landscape of Red Rock Coulee is created by badlands, hoodoos, and numerous huge, rust-colored boulders, called concretions. Many of these concretions are 2.5 metres across and are by far the largest, and best developed, in the province. Looking across the coulee, you can see concretions in many different stages of emergence as the softer shale of the Bearpaw Formation is slowly eroded away exposing the concretions. Some are still partly buried with only their tops visible, while others are completely exposed and scattered along the coulee walls.

### THE STORY

Concretions are a type of sedimentary structure that arouses a great deal of interest because of their widespread occurrence and variety of unusual shapes. They can show such remarkable symmetry that they have been erroneously mistaken for dinosaur eggs and brains, meteorites, and even fossil animals!

A concretion often begins as a dead organism, such as a leaf, bone or shell, that is buried in a soft "host" sediment such as sand or clay. The more homogeneous, or pure, the host sediment, the more favorable the conditions for large, spherical concretions. As subsurface groundwater trickles through the sediment, dissolved minerals are attracted to the fossil because it is chemically different from the host rock. The minerals accumulate and precipitate around the fossil, cementing together sediment grains until a dense rock structure, or

10

Site number and name of tourist destination region

★ 24 SOUTH

concretion, is formed. Concretions are harder than the surrounding sediments and therefore resist erosion. We know that iron oxide was one of the cementing minerals at Red Rock Coulee because of the striking red color of the concretions, which contrasts with the enclosing grey shale.

Once precipitation of minerals around the nucleus has begun, the concretion continues to grow outwards, producing a definite concentric layering, similar to an onion. This concentric structure is a key identifying characteristic of a concretion. Occasionally, a well preserved fossil can still be found in the centre of a broken concretion, although some concretions will form around an inorganic nucleus.

Concretions range in diameter from a few millimetres to several metres. We can only speculate why they were able to grow so large at Red Rock Coulee. There were likely long periods of uninterrupted growth with mineral-rich groundwater flowing continuously through the shale beds. In a few cases, where a concretion is only partly exposed, it is possible to see the shale beds are bent upwards over it and downward under it. This tells us that these concretions formed considerably after the Bearpaw shales were deposited. Red Rock Coulee has been designated as a Natural Area because of its beauty and fascinating rock forms.

Location map (background color indicates tourist destination region)

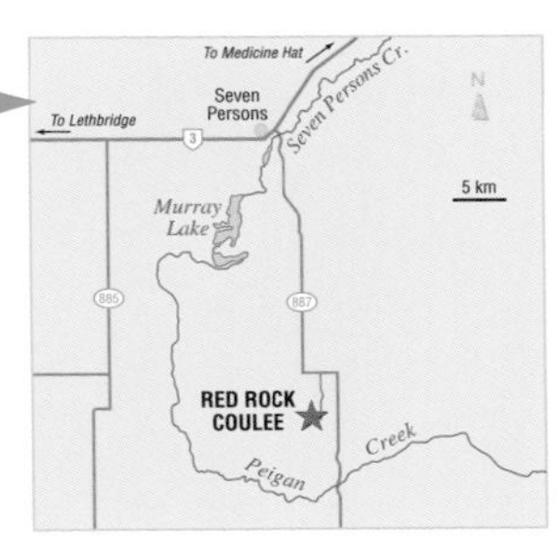


Interpretive line diagram

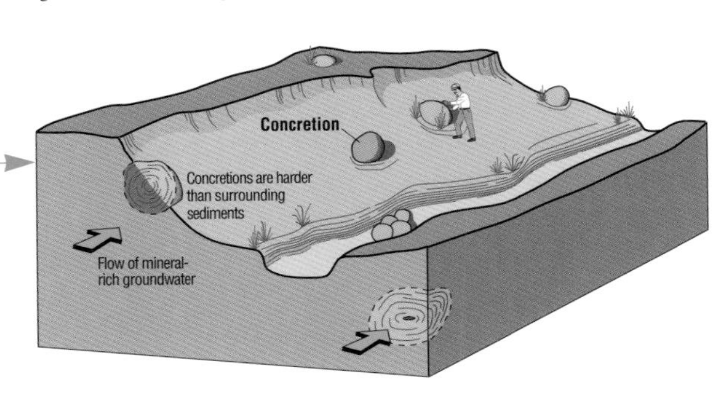


*Hard concretions weather out of soft shale bedrock.*

11

# A GEOLOGICAL PRIMER

You can enjoy Alberta's geological wonders without reading any of the material in this chapter. However, if you do read it, you will be rewarded by an increased understanding of the geology of the sites discussed in the book, as well as a greater appreciation of the geological richness of this province.

## THE LAY OF THE LAND

Geographically speaking, Alberta is divided into three major regions: Canadian Shield, Interior Plains, and Rocky Mountains and Foothills. Each region has its own distinctive landforms and geological structure. Alberta slopes gently downward from southwest to northeast from an elevation of 3747 metres, at the summit of Mt. Columbia in the Rockies, to less than 215 metres where the Slave River leaves Alberta on its way to Great Slave Lake.

## The Canadian Shield Region

The Canadian Shield is comprised primarily of hard igneous and metamorphic rocks, such as granites and gneisses. These rocks, of Precambrian age, represent the roots of ancient mountain ranges. Even though the Shield underlies all of Alberta, it is only exposed at the surface in the extreme northeast corner where it occupies about 15,540 square kilometres, or three per cent of the province. Hundreds of millions of years of erosion by water, and more recently by glaciers, has reduced these rocks to a relatively level, ice-scoured plain with polished and grooved rock surfaces. The eroded Shield rocks were carried away by advancing glaciers during the last Ice Age and we can now find them scattered across the rest of the province. In Alberta, the Canadian Shield has had little economic development but it is of great mining importance in other provinces.

C. Wallis – Natural Resources Service

*The Shield northeast of Fidler Point, Lake Athabasca. There are numerous lakes on the Shield because drainage is very poor — streams have not been able to erode valleys across the resistant, impermeable rocks. Lakes are elongated in one direction because they occupy ice-scoured rock structures such as ancient faults.*

C. Wallis – Natural Resources Service

*The Canadian Shield, near Tulip Lake, southeast of Ft. Smith. Much of the Shield's upland area is bare, ice-polished, pink rock where little vegetation can grow.*

## The Interior Plains Region

The Interior Plains is the vast expanse of land that lies between the Canadian Shield and the Foothills. It makes up 87 per cent of Alberta and has a great variety of landscapes including deeply carved river valleys, badlands, and high gravel-topped plateaus, such as Cypress Hills. The rocks that form the Plains vary in age from the arc of older Devonian limestones that borders the Shield to the younger Cretaceous and Tertiary sandstones and shales that extend westward to the Foothills. The rocks of the Interior Plains form a wedge that is six kilometres thick near its border with the Foothills and thins eastward until it reaches the Shield. Most of the surface of this wedge is hidden beneath a blanket of glacially deposited sediments, except along deep river valleys where the bedrock is often exposed as near-horizontal layers. The fertile soils developed on the glacial deposits form the basis of Alberta's agricultural industry. This rock wedge also contains a great deal of Alberta's petroleum, coal, and oil sands.

*Ron Mussieux – Provincial Museum of Alberta*

*Horsethief Canyon, near Drumheller. The Red Deer River has cut a deep valley into the Plains exposing the near flat-lying, soft sandstones, shales, and coal seams which rapidly develop into badlands. Here, the Plains are table-flat because they were once the floor of the former Glacial Lake Drumheller.*

*Alwynne Beaudoin – Provincial Museum of Alberta*

*Grande Prairie. These flat, former glacial lake beds form some of the richest farmland in Alberta.*

## The Rocky Mountains and Foothills Region

Although the Rocky Mountains and Foothills form only 10 per cent of the province, they are Alberta's most dramatic landscape and our most popular tourist destination. The icefields and glaciers that dot the landscape in the Rocky Mountains are remnants and reminders of Canada's glacial history. In the southern half of the province, the mountains tower high above the Interior Plains to the east and form Alberta's western boundary. Both the Rockies and Foothills are composed of the same sedimentary rocks as the Plains; however, in the Plains they are relatively flat-lying. In the mountains, the tremendous compression during mountain building has folded, faulted, and thrust the rocks eastwards forming a series of parallel, linear ridges. The Rockies in Alberta consist of three major belts — the Main Ranges, the Front Ranges, and the Foothills — each with its own distinctive rock types and physical appearance. The Main Ranges are Alberta's most westerly mountains and are formed of the oldest rocks, mainly quartzites, limestones, and dolostones. Generally, these mountains have the highest elevations. The Front Ranges are composed mainly of younger, more deformed limestones and dolostones. The Foothills have the lowest elevations. They are primarily made up of tilted or folded ridges of sandstones and shales of Mesozoic age.

*Ron Mussieux – Provincial Museum of Alberta*

*The Foothills southwest of Longview are composed of Cretaceous sandstone ridges and shale valleys. The topography is gentler and more subdued than the limestone Front Range Mountains to the west.*

*Ron Mussieux – Provincial Museum of Alberta*

*The intensely folded limestone of Mt. Michener is typical of the Front Ranges of the Rockies.*

# FORMATION OF ROCKS: Alberta's Foundation

Rocks make up the Earth's crust and are the most fundamental unit in geology. Each rock tells a story of how and where it was formed. They provide clues to the history of our planet; they record changes in the environment that occurred millions of years ago; and rocks containing fossils supply evidence of life forms over the last 3.5 billion years. As well, their economic importance is obvious: mining rocks gives us raw materials, such as iron, petroleum, coal, and gravel, that are essential in our everyday lives.

Geologists define a rock as a natural aggregation of one or more minerals. Some rocks do not have minerals but are made up of glasses (obsidian) or organic material (coal). There are three groups of rocks that are classified according to how they were formed: igneous, metamorphic, and sedimentary. Each type can be altered from one kind into another — this is part of the "rock cycle."

Metamorphic rocks form when pre-existing rocks, either igneous, sedimentary, or metamorphic, are altered by high temperature, high pressure, or both. Sedimentary rocks form when metamorphic, sedimentary, or igneous rocks are uplifted to the earth's surface where erosion then breaks them down into fragments, called sediments. When these sediments are transported, deposited and buried, they are compacted and cemented together to form sedimentary rocks. When sedimentary, metamorphic, or igneous rocks are buried deep beneath the earth's surface, they melt when the temperature gets hot enough. Igneous rocks are formed when molten rock, called magma, cools and hardens.

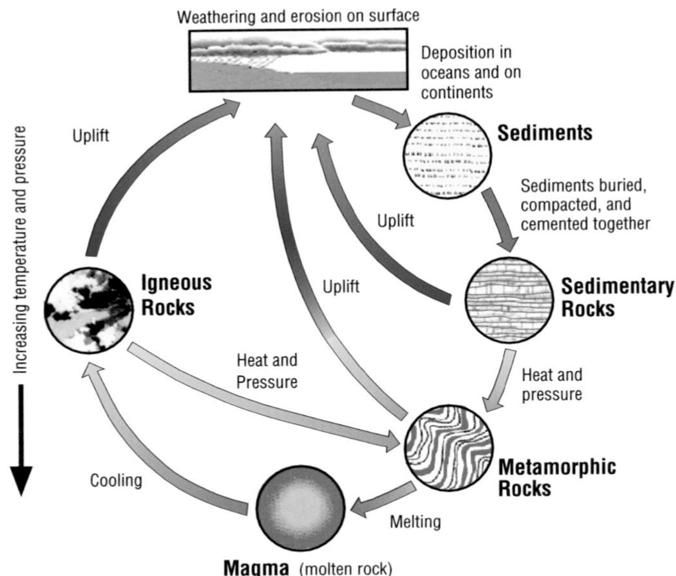

## Igneous Rocks

There are two groups of igneous rocks: those that hardened beneath the earth's surface are called intrusive or plutonic; those that hardened on the surface are called extrusive or volcanic. All igneous rocks are classified by the types of minerals present and by the size of their crystals. While the minerals reflect the chemistry of the original magma, the sizes and shapes of the crystals indicate how long it took for the magma to cool. Plutonic rocks, such as granite, have crystals large enough to be seen with the naked eye, indicating a slowly cooling magma within the earth's crust. Volcanic rocks, however, usually have microscopic crystals because the magma cooled very quickly when it was exposed to cool air on the surface. Basalt is the most abundant type of volcanic rock.

Most of Alberta's exposed bedrock is sedimentary but there are some outcrops of igneous rocks. Volcanic rocks can be found in the mountains of Waterton Lakes National Park and north to the Crowsnest Pass. Intrusive rocks, such as granites, form extensive parts of the Canadian Shield in northeastern Alberta. They can be seen particularly well at Fort Chipewyan and the Slave River Rapids. There are also small outcrops of such rocks south of the Milk River.

Even though outcrops of igneous rocks are uncommon in Alberta, the advancing continental glaciers during the last Ice Age plucked blocks of both igneous and metamorphic rocks from the Shield and scattered them across most of the province. Thus, such rocks are abundant in gravel pits, gravel bars, and rock piles in farmer's fields.

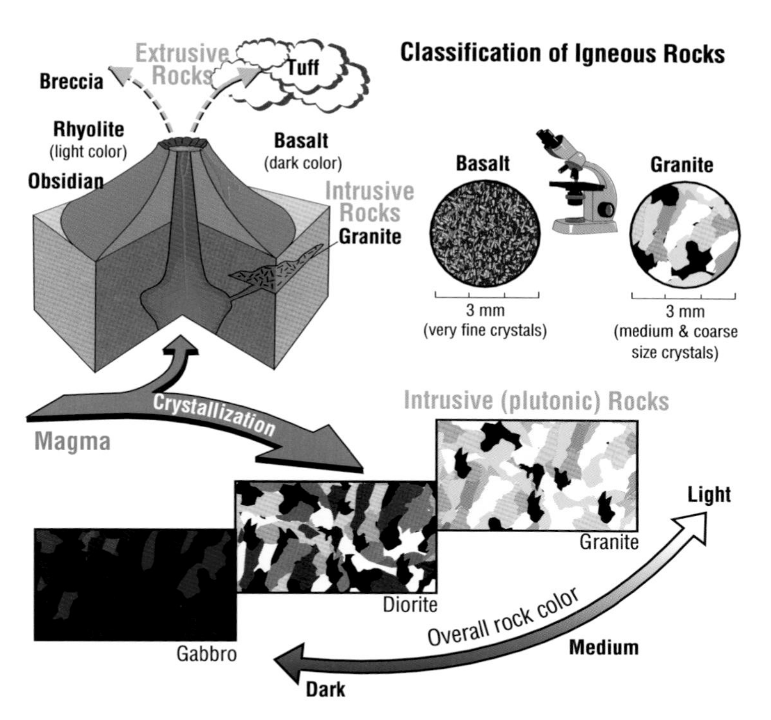

## Metamorphic Rocks

Under intense pressures and temperatures deep within the earth, the minerals in a rock can be altered, without melting, into new minerals. Combinations of these new minerals form metamorphic, or "changed," rocks. Metamorphic rocks can be formed throughout large regions, for example during mountain building. Or they can occur within small areas, such as the metamorphoses of rocks that surround a molten rock mass. Many metamorphic rocks are banded because of the rearrangement of new minerals. In gneiss, for example, the banded texture is caused by the segregation of dark and light minerals. These dark and light layers can often be irregular, particularly when the rock has been folded under pressure. In metamorphic rocks, such as schist, minerals are often plate-like and parallel, creating a texture called foliated.

Metamorphic rocks form a small percentage of Alberta's outcrops and can be found only in isolated localities around Alberta. The Canadian Shield is the best area to find high grade metamorphic

**Classification of Metamorphic Rocks**

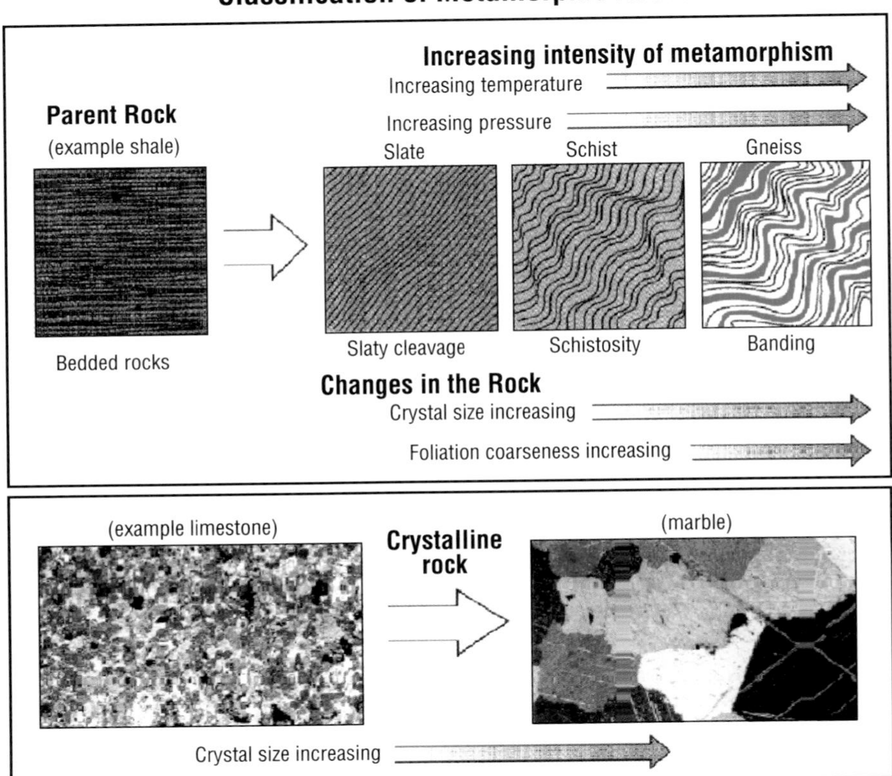


*With increasing intensity of metamorphism, there is a transition from one rock into another one, each with distinct minerals and textures. For example, as heat and pressure increase, shale, a sedimentary rock, slowly turns into slate, then schist, and finally gneiss.*

rocks, such as gneiss, which is particularly visible at Pelican Rapids (see page 54) and Mountain Rapids (see page 52). High grade metamorphic rocks are those that have been subjected to extremely high pressures and temperatures during their formation. Quartzite is another common metamorphic rock in Alberta and is one of the major rocks in the Main Ranges of the Rocky Mountains. Between Waterton and Crowsnest Pass, and in the mountains near Jasper, there are numerous outcrops of low grade metamorphic rocks. Low grade metamorphic rocks are those that have been subjected to relatively low pressures and temperatures during their formation.

Like igneous rocks, metamorphic rocks were also strewn about the province by glaciers during the last Ice Age, and are also commonly found in gravel pits, river gravel bars, and rock piles in farmer's fields.

## Sedimentary Rocks

Sedimentary rocks cover about three-quarters of the earth's surface and form more than 90 per cent of Alberta's bedrock surface. Sedimentary rocks can be made up of rock fragments, called sediments, that have been eroded from one place and moved to another by water, ice, wind, or gravity. In time, the sediments are buried, compacted and cemented together to form sedimentary rocks. Rocks that are made up of sediments are called "clastic" sedimentary rocks, and include conglomerate, sandstone, and shale. Other sedimentary rocks are of chemical or organic origin. Chemical sedimentary rocks are formed by the precipitation or evaporation of minerals from solution in ancient seawater, and examples from Alberta include limestone, gypsum and halite (rock salt), and dolostone. Even though dolostone can be formed by precipitation, it is usually formed when limestone is altered by the addition of a magnesium-rich solution. Organic sedimentary rocks are formed by the accumulation of dead plant and animal matter which is then compacted together.

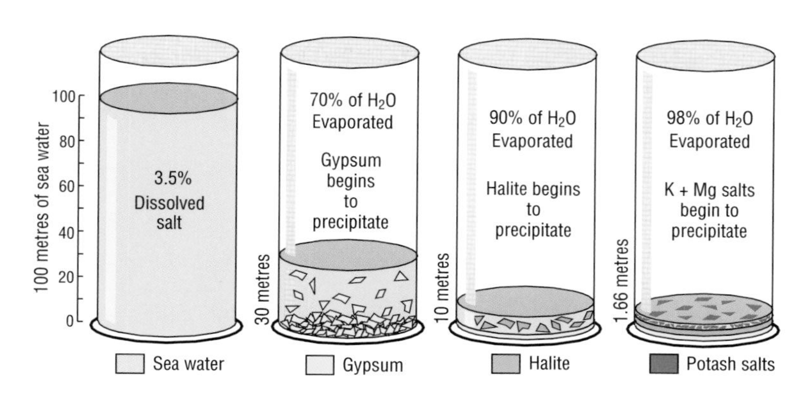


*Some sedimentary rocks form by the evaporation of sea water.*

## Classification of Sedimentary Rocks

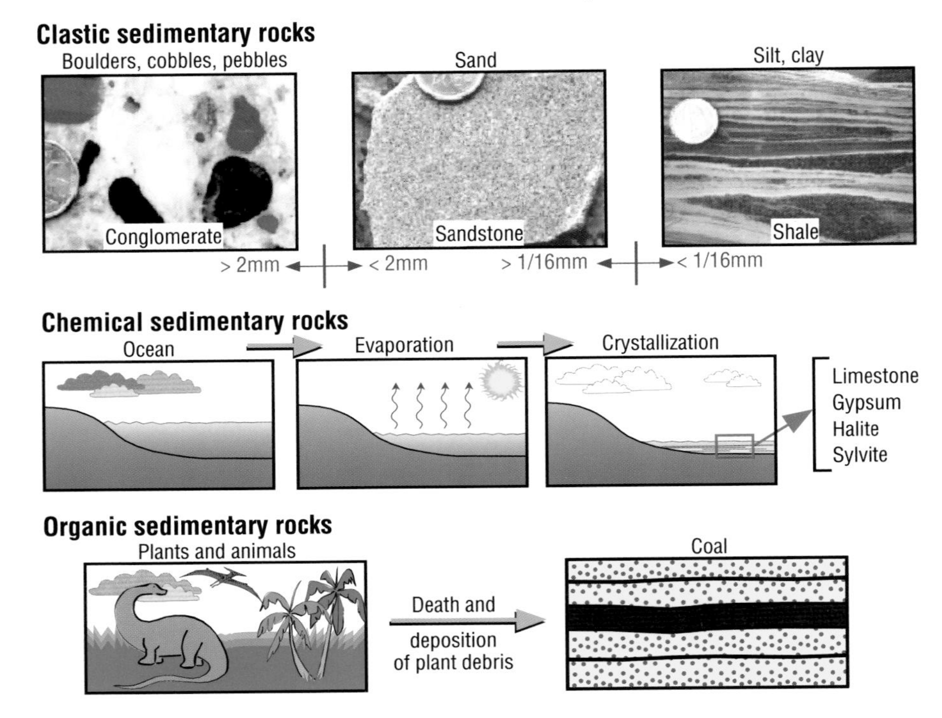


All three kinds of sedimentary rocks are found in Alberta. The Rocky Mountains consist mostly of the chemical rocks limestone and dolostone, with some sandstones and shales. The Plains are mainly sandstone and shale, although there is some gypsum, limestone, and dolostone forming a rim around the Canadian Shield in northeast Alberta. The organic rock coal is found in both the Plains and Rocky Mountains.

# SOME OF ALBERTA'S SPECIAL ROCKS: Caves and Meteorites

*Rob Young – University of Alberta*

*Quaternary paleontologist Jim Burns crawls carefully under "soda straw" stalactites. These features grow downwards from the ceiling of the cave when calcium carbonate is precipitated by dripping water.*

## Caves

Alberta is the home of some of Canada's largest and deepest caves. Castleguard Cave, in Banff National Park, is Alberta's longest explored cave, with a length of just over 20 kilometres. All Alberta caves are formed where groundwater dissolves the subsurface soluble bedrock. Ice caves are also formed in this manner but at the present time their rock temperature remains below freezing most of the year and a build-up of permanent ice occurs in the caves. Large, beautiful hexagonal ice crystals can grow from the cave walls.

The geologic definition of a cave is an underground open space, generally with a connection to the surface and large enough for a person to enter. The cave explorer's definition requires that an entrance lead to some bedrock passage, and that the cave be formed by the dissolving of bedrock rather than by frost shattering or breakdown. This dissolution generally occurs where the bedrock has been below the water table or springs have been present.

For caves to be formed, there must be groundwater trickling through subsurface bedrock — either in carbonates, such as limestones and dolostones, or in gypsum. In Alberta, most caves are formed in carbonates of the southern Rockies and Foothills, while a few can be found in the gypsum of Wood Buffalo National Park. The development of gypsum caves, although rare, is quite simple because it is a very soluble rock and dissolves quickly when groundwater contacts it. Cave development in the Rockies and Foothills, however, is more complicated because, in order for the carbonate bedrock to be dissolved, groundwater needs to be somewhat acidic. Groundwater is made acidic when carbon dioxide from the atmosphere and soil combine with the water to form carbonic acid.

Structures seen inside a cave are formed by the accumulation of calcium carbonate minerals, such as calcite, from dripping water. The most common cave structures are simple coatings and crusts, but there are some well preserved, although very fragile, stalactites and stalagmites. Stalactites are icicle-like formations that hang from the roof of the cave while stalagmites are pinnacles that build up from the floor. Both can meet to form impressive columns from roof to floor. These structures are produced when carbon dioxide-rich water trickles down through cracks in the limestone and dissolves the rock. As this water enters a cave, the carbon dioxide comes out of solution and the calcium carbonate is precipitated.

*Natural Resources Service*

*Beautiful hexagonal ice crystals growing on the walls of the Plateau Mountain ice cave.*

Many animals, such as pack rats, bats, snakes, insects, and fish have adapted to a cave environment. It should be noted that all caves, especially gypsum caves, are dangerous and should only be entered with an experienced guide. If you are planning on visiting a mountain cave, it is best to contact Parks Canada or the Alberta Speleological Society.

## Meteorites: Extraterrestrial Rocks

Provincial Museum of Alberta

*Iron Creek meteorite.*

Meteorites are rocks that have travelled from space and landed on the surface of the Earth. They are the oldest rocks known on our planet and are therefore an invaluable source of information about the geology of the solar system and this planet. Geologists believe that meteorites and Earth formed at the same time. Meteorites have been dated as being about 4.5 billion years old. Therefore, this is a good estimate of the age of our planet.

Meteorites probably originated when asteroids or comets collided in space and broke into fragments. These fragments eventually hurtled through the earth's atmosphere where they were heated by the immense friction that developed. Those that survived the trip through the atmosphere and collided with Earth are called meteorites, and those that burn up completely are called meteors.

Meteorites are divided into three groups: stony meteorites, iron meteorites, and the rare stony-irons. Meteorites have a wide range of sizes and appearances. In spite of their similarities to man-made materials and natural rocks, they have a number of distinguishing characteristics that will help you recognize them. Because of the iron in them, they are magnetic and rusted, even on broken surfaces. They usually have a thin, dull black or brown crust which was formed as the surface was melted in the atmosphere. Meteorites tend to have irregular shapes, are quite heavy, and often have pitted surfaces.

So far, fourteen meteorites have been found in Alberta, more than any other province, probably because extensive cultivation uncovers buried meteorites. Two of Alberta's best known ones, the Bruderheim and the Iron Creek, are discussed here.

The stony Bruderheim meteorite is composed mainly of silicate minerals. Even though stony meteorites are the most common type, they are difficult to recognize because they resemble "ordinary" rocks. Bruderheim meteorite fragments fell over an area of several kilometres just north of Bruderheim, Alberta, on March 4, 1960 at 1:06 a.m. It entered the atmosphere at a velocity of six kilometres per second and exploded. The resulting flash was visible 300 kilometres away and the explosion could be heard over 100 kilometres away. Nearly 700 fragments with a total weight of 303 kilograms have been found, making it Canada's largest known meteorite. Most of the fragments are stored at the University of Alberta but a few are on display at the Provincial Museum of Alberta.

The Iron Creek meteorite, an iron meteorite of mostly iron with some nickel, weighs 175 kilograms. Iron meteorites are dense, heavy and strongly magnetic. Geologists think that the Earth's interior is made of materials similar to iron meteorites — this would account for the high density of the earth. The Iron Creek meteorite fell long before Europeans arrived on the Prairies. It was found on top of a hill near Iron Creek, close to the town of Sedgewick. This meteorite is now on display in the Syncrude Gallery of Aboriginal Culture at the Provincial Museum of Alberta.

*Provincial Museum of Alberta*

*Bruderheim meteorite.*

# THE GEOLOGIC TIME SCALE

Scientists believe the earth to be about 4.5 billion years old — an almost incomprehensible amount of time. Geologists have separated this time into eons, eras, and periods, each with its own name. Periods are the most basic unit of geologic time. Many of their names are derived from Western Europe where much of the initial time scale development was done.

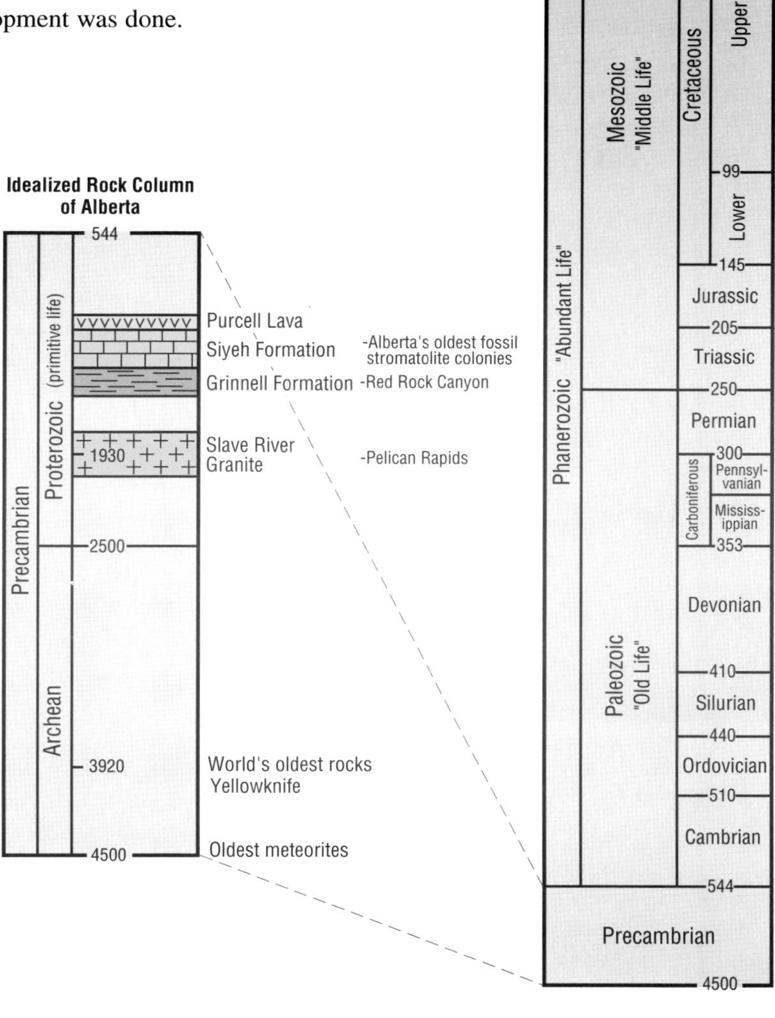

Ron Mussieux – Provincial Museum of Alberta

*Drilling rig, Leduc #1. A 1940-1950s vintage drilling rig has been erected at the Leduc #1 Wellsite (see page 68).*

Rocks representing most of these time periods are found in Alberta. The best way for geologists to see subsurface rocks in their proper sequence is to look at oil well drill cores. Normally, the well starts in younger rocks and cuts through successively older rocks, and if one continued drilling, the Precambrian basement of Alberta would eventually be reached.

During the drilling process, rock fragments are brought to the surface and identified by a geologist. These rocks are used to produce a diagram, called a rock column, that shows rock types, thicknesses, and fossils. Related rock layers are grouped into formations and given a name. Formations are often combined into larger entities called groups. Formations are the basic rock unit that geologists use when they make geological maps.

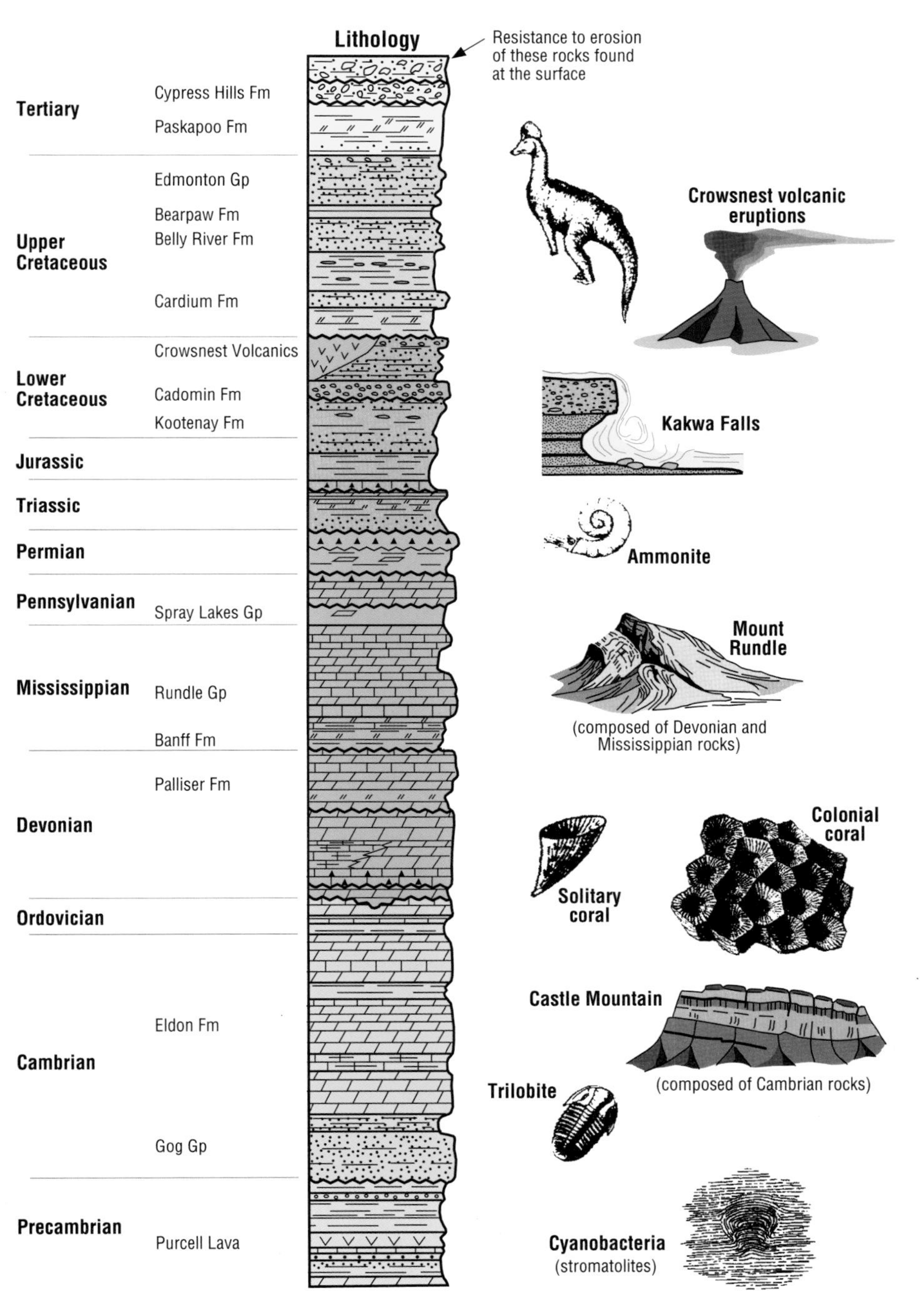


## An Idealized Rock Column for Alberta

# RICHES IN ALBERTA'S ROCKS AND MINERALS

Rocks and minerals are fundamental to Alberta's economy. In some parts of the world, people have used rocks and minerals for weapons, tools and adornment for over a million years, and in Alberta, stone artifacts have been discovered that are over 11,000 years old. The demand for rock and mineral resources is greater today than ever before; they are necessary to maintain our highly advanced and technological civilization. Alberta is underlain by thousands of metres of rocks which hold many valuable resources and, in fact, the resource production in this province is more than half of Canada's total production. The vast majority of Alberta's mineral industry consists of energy resources, or fossil fuels, including coal, oil, gas, and gas byproducts, such as sulphur. The province is also a significant producer of cement, sand and gravel, lime, salt, and gold. Therefore, the study of rocks is more than an academic pursuit, and Alberta has justifiably earned the title of "Canada's energy province."

**Alberta's Mineral Production (1996)**
**$26 billion**

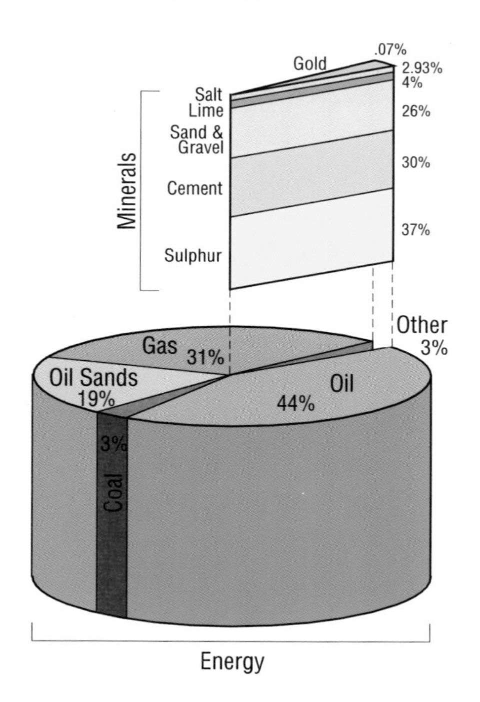


**Canada's Mineral Production (1996)**
**$49.2 billion**

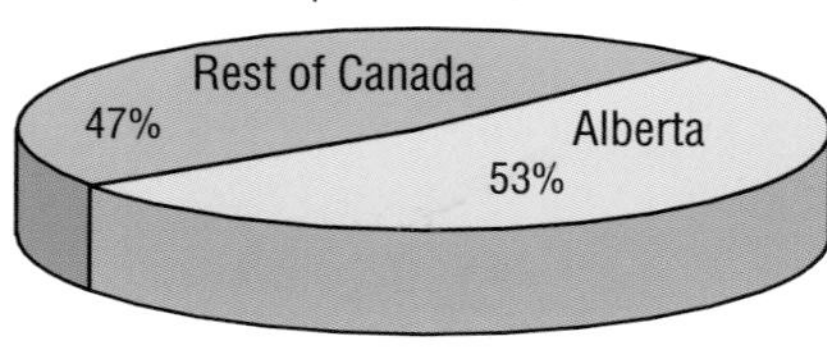


## A Geological Cross Section of Alberta

Another common diagram used by geologists is the cross section. In order to envision this type of diagram, imagine that part of the earth's crust is sliced like a great pie and a piece is lifted out so you can look at the rock layers on their edge. The information needed to make a cross section, such as depth, rock types, and thicknesses of rock layers, is obtained from the rock columns of several wells. Wells are selected along a straight line and the rock layers represented on the columns of several wells are joined together to produce a view of a slice through the crust.

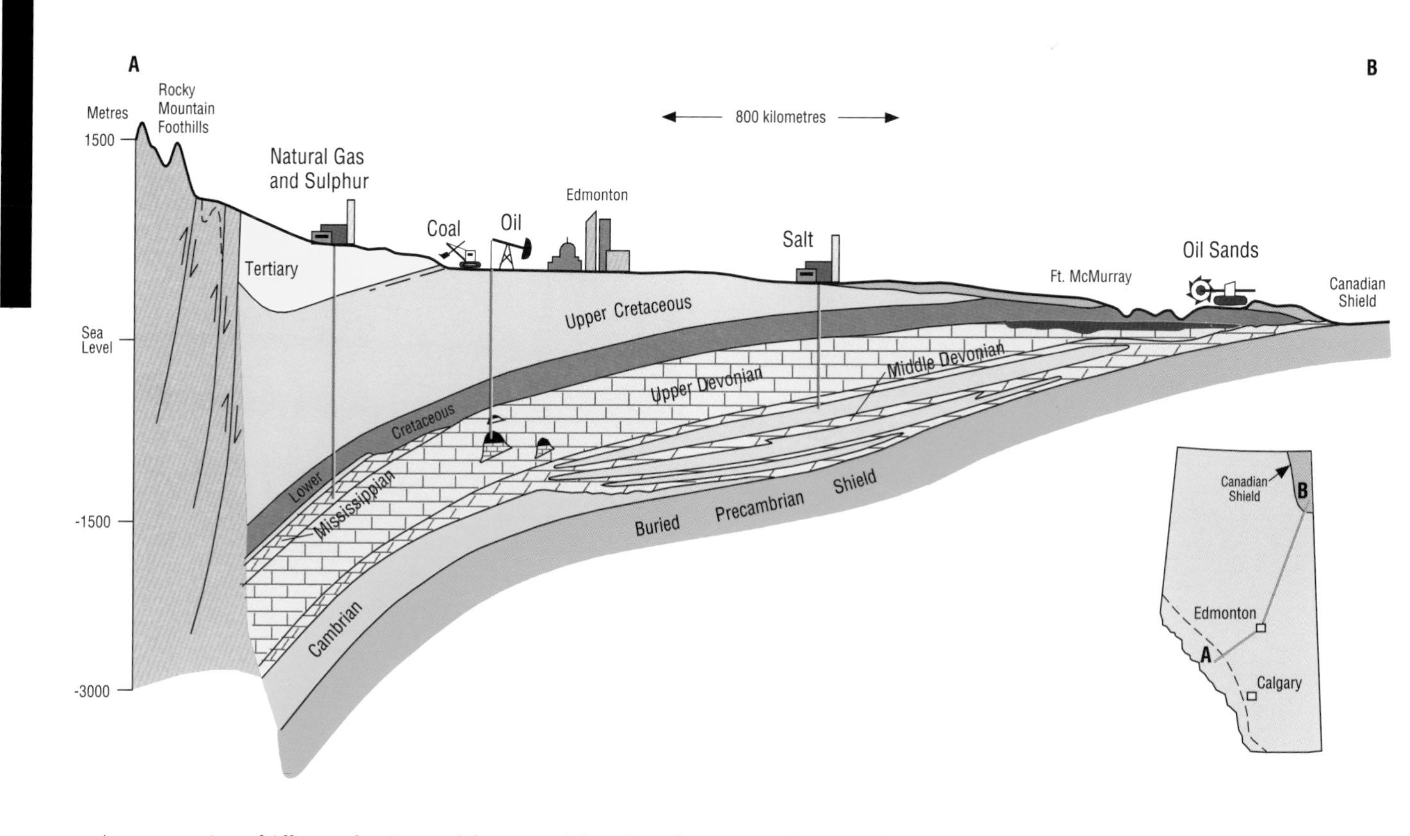


*A cross section of Alberta showing rock layers and the mineral resources they contain.*

# HOW THE ROCKIES WERE RAISED

Geologists are often asked, "How old are the Rockies?" The response is usually another question, "Do you mean how old are the rocks?" or "When were the rocks thrust up into mountains?" or "When did the Rockies begin to look like they do today?" The building of the Rocky Mountains can be viewed very simply in three major phases: deposition of the rocks; uplift of the rocks into mountains; and a long period of erosion and sculpting by water and glacial ice.

## Plate Tectonics - The Mechanism

The earth's crust is constantly changing. It is made up of 16, rigid interlocking rock slabs, called plates, that float on top of a hot, semi-fluid layer. At the boundary between some plates, called mid-oceanic ridges, hot molten rock, or magma, wells up from below and pushes the plates apart. They move very slowly, often for millions of years, before they collide with, slide past, or are pushed under another plate. These plate interactions are geologically violent events that result in earthquakes, volcanic activity, and mountain building. As a result, crust is either being added, destroyed, or altered where plates meet. Recent, and well known examples of plate interactions include the explosive eruption of Mt. St. Helens in Washington and earthquakes along the San Andreas Fault in California.

*Ron Mussieux – The Provincial Museum of Alberta*

*Limestones and shales that were originally deposited as horizontl beds are deformed into complex folds during mountain building. Mt. Michener.*

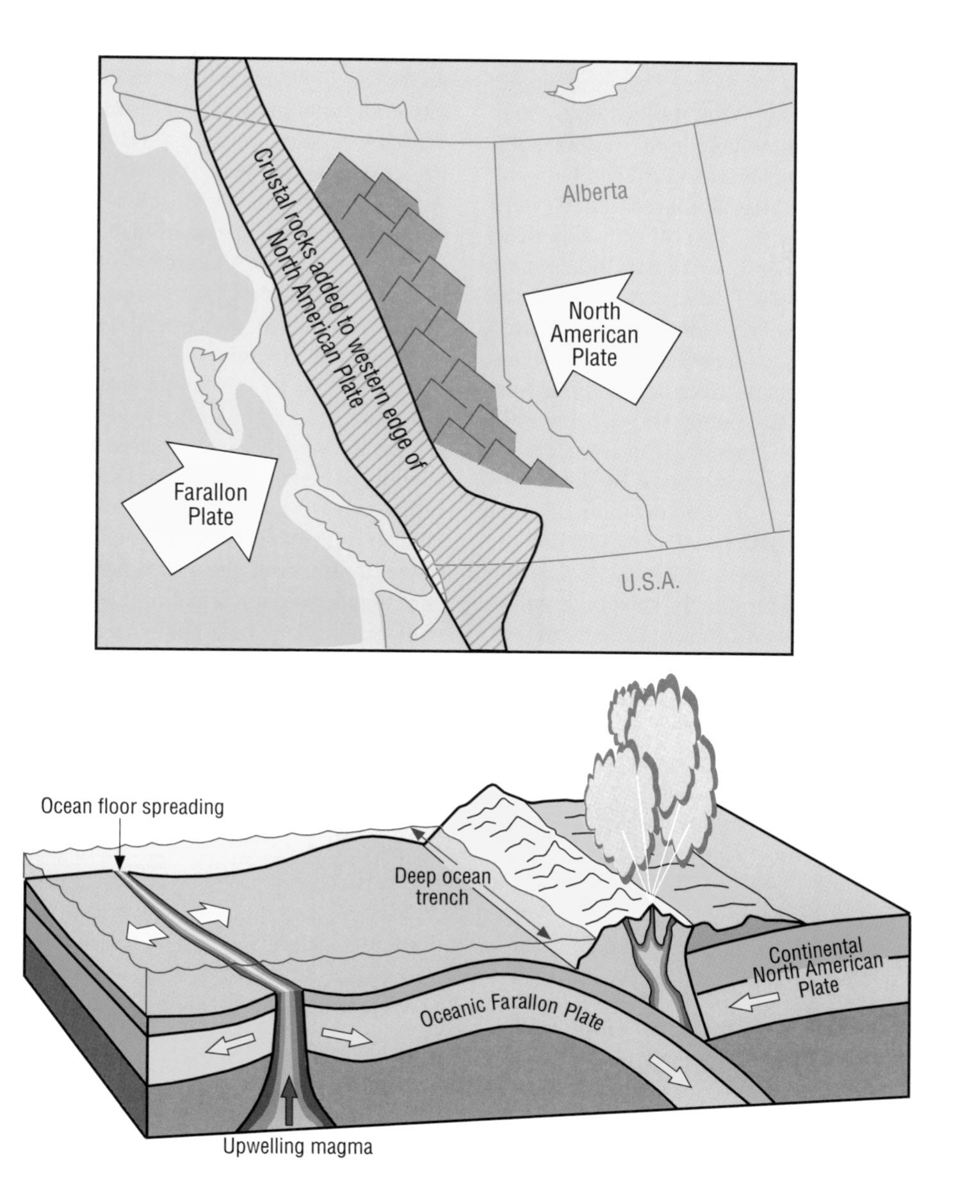


**Plate Tectonics - The Mechanism**

## PHASE I: Deposition (1500 to 170 million years ago)

The Rocky Mountains were once near-horizontal layers of marine sediments. These had accumulated on the ocean shelf that bordered the western edge of the ancient North American continent. For over two billion years, these sediments were eroded off the exposed Shield to the east, transported by rivers, and deposited in the ocean. There was also a great deal of limestone and dolostone being formed as well, either by the chemical precipitation of lime from the seawater, or by the organic deposition of lime by animals and plants, in the form of reefs. Under the weight of the accumulating sediments, the ocean floor slowly subsided, which in turn allowed more sediments to collect. Eventually, all of these layers were compressed and cemented into sedimentary rocks, such as sandstone, shale, limestone, and dolostone. This quiet time of deposition in the ocean continued until the middle of the Jurassic Period, about 170 million years ago. At this time, "the big crunch" began and the Rocky Mountains began their rise from the depths of the ocean.

## PHASE II: The Big Crunch and More Rock Deposition (170 to 55 million years ago)

In the middle of the Jurassic Period, fragments of the earth's crust drifted northwards where they collided with, and became welded onto, the North American Plate. These are the blocks of crust that make up much of British Columbia. The compression from these collisions folded, faulted, and shoved up the ocean floor sediments that had been deposited in Phase I. The rocks broke into enormous slabs, called thrust sheets. With successive collisions and continuing compression from the west, sheets were pushed eastwards, along thrust faults. Each sheet slid up over the one in front of it, gradually creating a symmetry similar to shingles on a roof. These sheets formed the ancestral mountain ranges of the Rockies.

In response to this compression and mountain building, the rocks on the east side of the new mountains were warped downwards creating a basin, essentially in the area of the Interior Plains today. When this happened, ocean waters flowed from both the north and south into the depression creating an inland seaway across the continent. For the next one hundred million years, sediments were eroded from both the mountains to the west and the Shield to the east and deposited in this sea. These sediments were even thicker than those that had been deposited on the ocean floor, and the sea became filled with sediments. Extensive deltas, swamps, and forests developed, and in these environments, dinosaurs flourished. Then, during the Late Cretaceous Period, there was another violent episode of plate collisions to the west. This caused another period of compression and the resulting uplift of the inland sea sediments and mountain building. The existing mountains were thrust up even higher and, with increased compression, first the Main Ranges were formed, then the Front Ranges, and finally the Foothills. By the end of this mountain building, the original sediments were stacked on top of one another shortening their length by 200 kilometres!

## PHASE I: Deposition

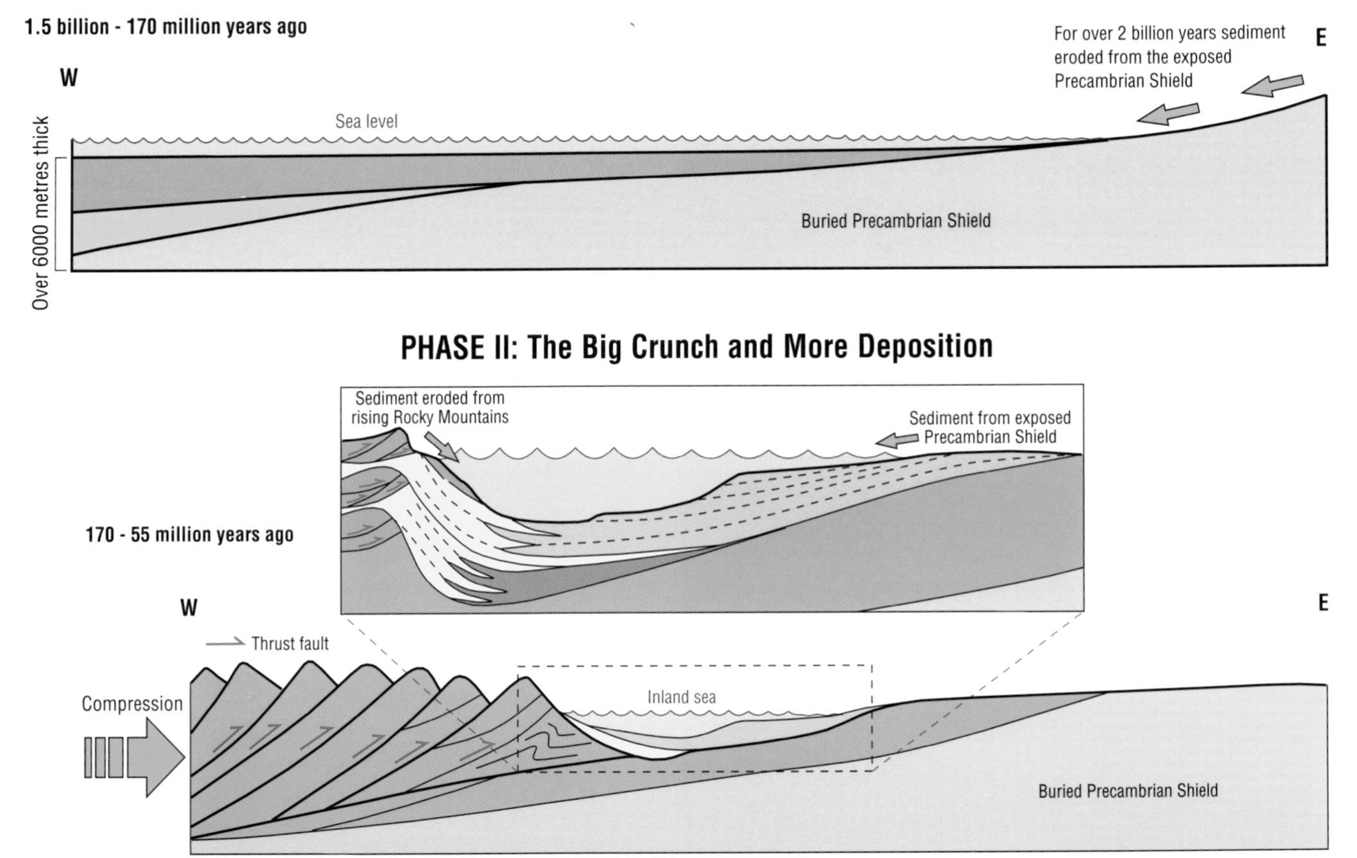

## THE END RESULT OF MOUNTAIN BUILDING: A Cross Section Through the Rocky Mountains to the Interior Plains

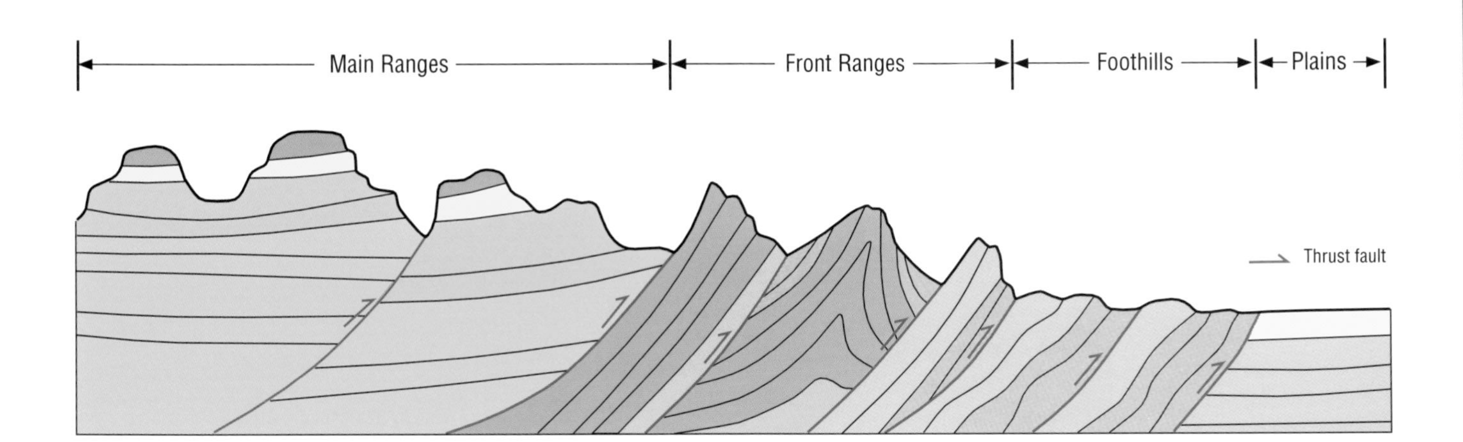


*The rock layers of the Main Ranges are relatively flat-lying and less deformed that those of the Front Ranges and Foothills. The Front Ranges are composed mainly of steeply dipping, or tilting, layers of limestones and dolostones of Paleozoic age, while the Foothills are composed of Mesozoic-aged sandstones, shale, and coal.*

## PHASE III: Sculpting the Face of the Rockies (55 million years ago to the present)

Although erosion of the mountains has been ongoing since they first rose above sea level, it was always exceeded by the rate of mountain building. By 50 million years ago, though, mountain building ceased and erosion by water and ice gained the upper hand. The most striking changes in the landscape have occurred in the last 1.65 million years with the onset of global cooling and the advance of the continental and mountain glaciers. The glaciers scoured and polished the land and left behind thick layers of sediments that had been carried in the ice.

### Mountain Ranges Before Glaciation

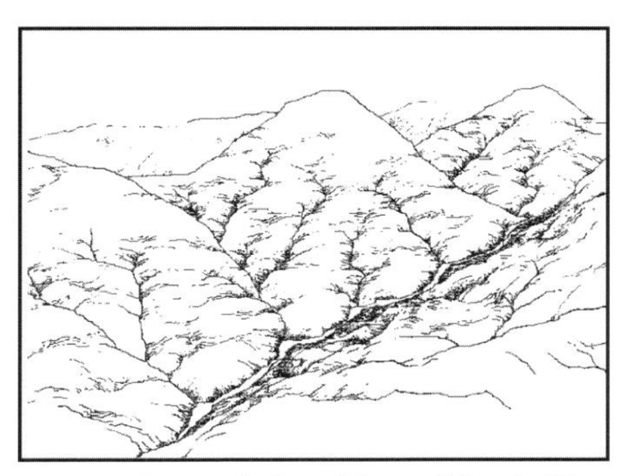

Courtesy Geological Survey of Canada, Misc. Report 26, 1976.

*Notice the rounded summits and V-shaped valleys. This well developed drainage system had many little creeks, and few waterfalls and lakes.*

### The Ice Age

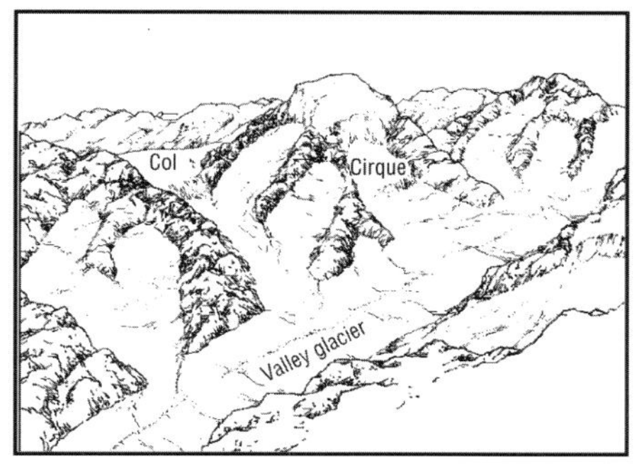


Courtesy Geological Survey of Canada, Misc. Report 26, 1976.

*Snow accumulated and built up into thick ice sheets and glaciers. They flowed downslope and slowly joined together to fill old river valleys. Glaciers are an extremely powerful force of erosion.*

### The Mountains Today

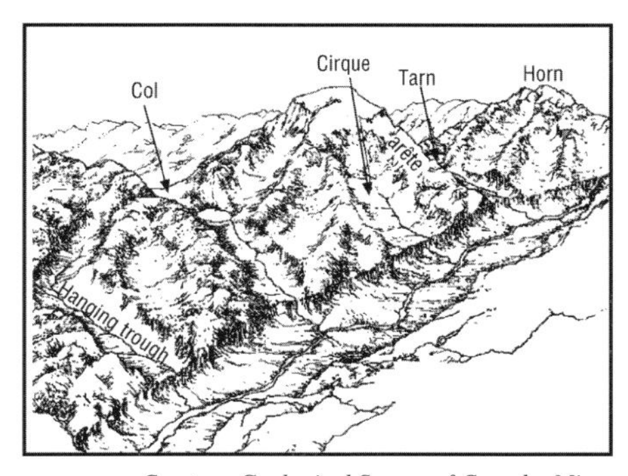


Courtesy Geological Survey of Canada, Misc. Report 26, 1976.

*Following the melting of the glaciers, a dramatically different landscape appeared. These are the modern Rocky Mountains. Glaciers have carved out broad, U-shaped valleys with flat bottoms and steep walls. On the mountain sides are scooped out depressions, called cirques, which often contain tiny tarn lakes. Sharp rock ridges, called arêtes, are formed by the walls of adjacent cirques that often still contain small glaciers.*

# CHAPTER 2: ALBERTA'S NORTH

*The Canadian Shield, north of Lake Athabasca, northeastern Alberta.*
*C. Wallis – Natural Resources Service*

## ALBERTA'S NORTH

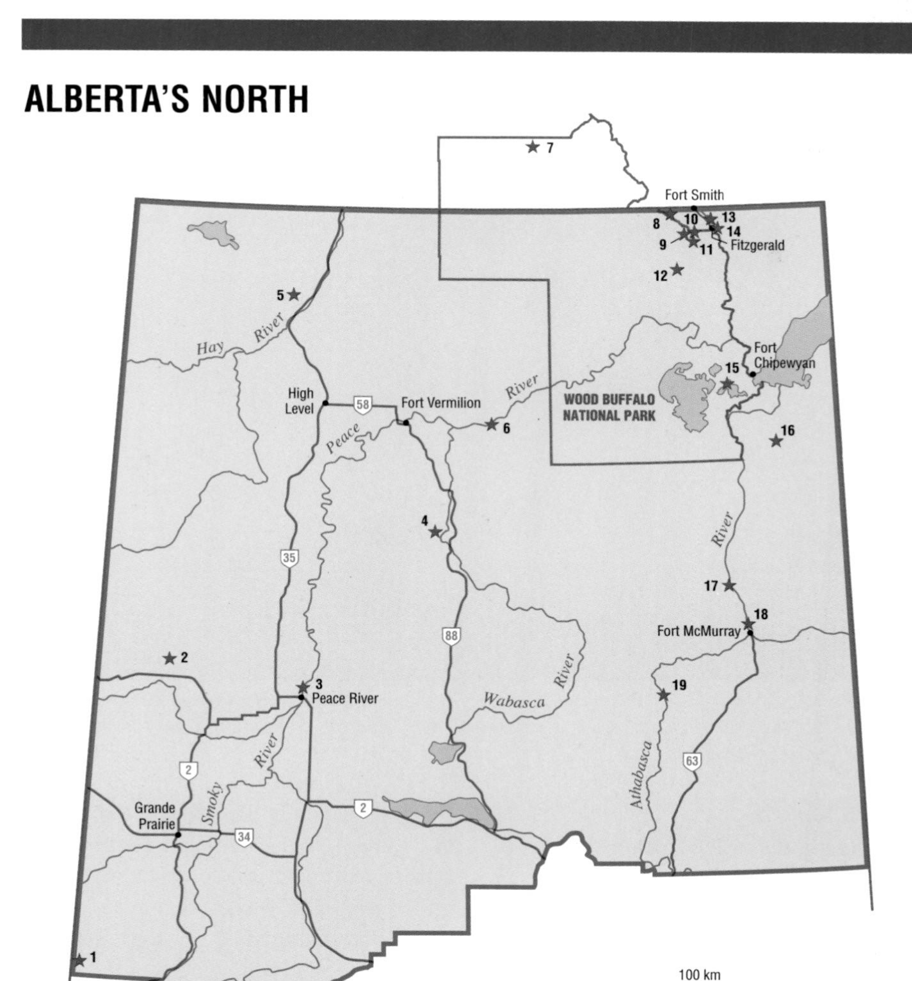


★ **1** Kakwa Falls
★ **2** Clear Hills Iron Ore
★ **3** Peace River Landslides
★ **4** Burning Sulphur
★ **5** Hot Pot
★ **6** Vermilion Chutes
★ **7** Wood Buffalo National Park Angus Fire Tower Sinkhole
★ **8** Wood Buffalo National Park Salt Plains Lookout
★ **9** Wood Buffalo National Park Karstland Trail
★ **10** Wood Buffalo National Park Domed Rocks
★ **11** Wood Buffalo National Park Salt-etched Rocks
★ **12** Wood Buffalo National Park Pine Lake
★ **13** Mountain Rapids
★ **14** Pelican Rapids
★ **15** Peace-Athabasca Delta
★ **16** Athabasca Dunes
★ **17** Athabasca Oil Sands
★ **18** Athabasca River Valley
★ **19** Grand Rapids

# KAKWA FALLS: A Scenic Waterfall of the Northern Foothills

Don Taylor – Provincial Museum of Alberta

*Kakwa Falls is formed over a ledge of Cadomin conglomerate. Notice the cave formed in softer rock behind the falls.*

## HIGHLIGHTS

Kakwa Falls is the main attraction of the Kakwa Wildland Provincial Park which is in the Foothills, 150 kilometres southwest of Grande Prairie. Besides being a place of great beauty, it also displays an outcrop of the Cadomin Formation, one of the most prominent ridge and ledge-forming rock deposits in the area. The last portion of the road to the falls is passable only by 4-wheel drive vehicles.

## THE STORY

The Kakwa River, fed by Lake Kakwa, is a tributary of the Smoky River. The falls are formed as the river flows over a large anticline, or folded arch of bedrock. Looking downstream from the lookout at the falls, you can see that the prominent Cadomin Formation forming the falls also forms the lip of the canyon. This demonstrates that the falls have moved slowly upstream as the underlying rock was eroded away.

The river plunges a vertical distance of approximately 30 metres. Like most falls, they are the result of a river crossing an outcrop of resistant rock which overlies softer, and more easily eroded rock. Here, the erosion-resistant rock is the Cadomin Formation which was deposited around 118 million years ago. The Cadomin Formation is a conglomerate, which is often referred to as "nature's concrete," that has thin layers of coal and sandstone. It is made up of rounded pebbles, cobbles, and boulders of multicolored chert and quartzite, with a matrix of well cemented sand. These sediments were deposited as alluvial fans and braided river deposits west of a narrow river channel that once flowed parallel to the then rising Rocky Mountains. As ancient rivers flowed off the steep mountains and into the channel, the break in slope caused the rivers to dump their load of gravel and sand. Around the Kakwa River area, the Cadomin Formation is nearly 26 metres thick, and farther west it reaches thicknesses of 200 metres! This formation is one of the more recognizable rock formations in the Foothills and is an important marker for subsurface well logs in oil exploration. During early petroleum exploration, the Cadomin proved to be a major problem for drilling in the Alberta Foothills because its extreme hardness quickly wore out drill bits.

The distinctive feature of these falls is the degree of undercutting that has taken place. There is a large cave behind the falls formed by water splashing and spraying against the back wall of the falls and eroding it. The boundary, or contact, between the upper resistant layer and lower weak layer is what forms the ceiling of the cave.

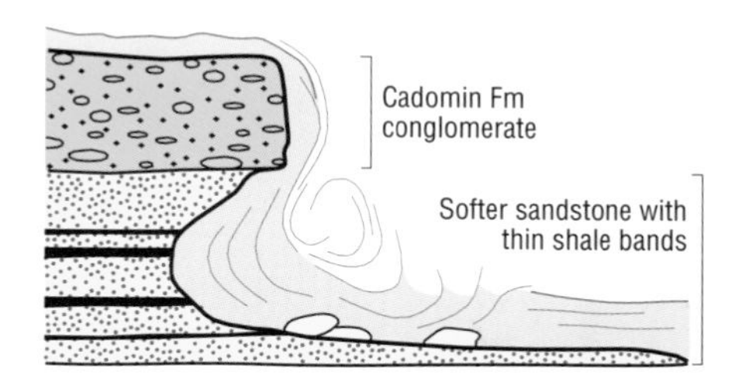

Ron Mussieux – Provincial Museum of Alberta

*The steep canyon walls downstream of Kakwa Falls are formed of the same hard Cadomin conglomerate.*

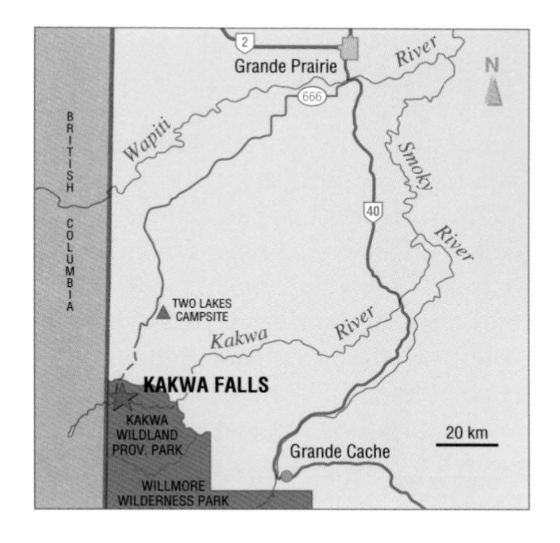

# CLEAR HILLS IRON ORE DEPOSITS: A Potential Alberta Iron Mine

*Ron Mussieux – Provincial Museum of Alberta*

*Geologist standing in the test pit of brightly colored iron-rich sandstone, Clear Hills.*

## HIGHLIGHTS

Alberta is Canada's energy province, but it still does not have a viable metal-mining industry. Does it have the potential to develop its own steel industry? A growing demand for iron and steel products during the 1960s led to widespread interest in an Alberta source of iron ore. On the basis of grade and volume, the iron-rich sandstones in the Clear Hills west of Peace River contain the largest potential source of such ore in the four western provinces. These rusty-red sandstones are highly visible along the secondary road leading to Running Lake campsite, north of Worsley.

## THE STORY

These deposits were originally staked by prospectors in 1925 who called it a "mountain of iron." Serious investigation and testing for economic potential did not occur until after 1953, however, when the sandstones were rediscovered during oil and gas drilling. By 1957, Premier Steel Mills Ltd. had developed an interest in the area and stripped off enough overburden to remove a 50 tonne sample from a test pit. Excitement ran high when the analyses showed an iron content of between 29-35 per cent, with estimated total reserves of 1.5 billion tonnes of iron ore.

These figures generated such interest that Premier, in co-operation with the Alberta Research Council, sponsored a pilot-plant test in the United States to determine the ore's economic feasibility. In addition, it was also determined that, unlike any iron and steel works in eastern Canada, Alberta could obtain all the raw materials needed for iron ore

Ron Mussieux – Provincial Museum of Alberta

*View from the top of Clear Hills southwards towards the plains, almost 800 metres below.*

mining from within the province. Despite all of this, a local steel industry has still not developed here because of the high costs of removing impurities from the deposit. As well, the iron-rich layer of sandstone runs horizontally through Clear Hills and any strip mining would have to remove the top from these hills completely, including its forests; as a result, development would be a significant environmental issue.

The original test pit is the best place to see these brilliantly colored sandstones. The sand grains were originally deposited about 85 million years ago as a series of sand bars in the Late Cretaceous sea that covered much of Alberta. Each grain was rolled back and forth by the currents and waves, and gradually layers of iron oxides accumulated around it. Close examination of the sandstone reveals it is actually made up of spheres, resembling tiny fish eggs. Geologists call these spheres "oolites." Surprisingly, although this is a marine deposit, many wood fragments are incorporated into the sandstone. These were probably washed offshore and became lodged in the accumulating sand bar. The potential to develop a Clear Hills iron mine remains, but it is not likely to happen in the near future.

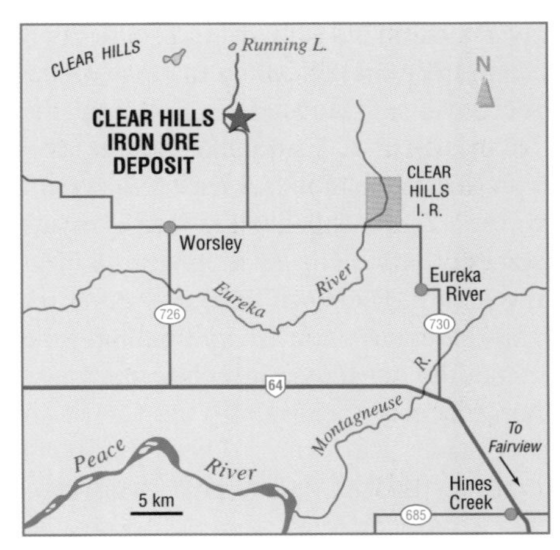


*Test pit is located 10 kilometres north of the T-intersection, east of Worsley.*

# PEACE RIVER LANDSLIDES

*Ron Mussieux – Provincial Museum of Alberta*

*View from Sagitawa Lookout on Judah Hill Road toward the town of Peace River showing numerous landslides that have occurred on weak shale bedrock.*

## HIGHLIGHTS

From its source in British Columbia, the Peace River winds across northern Alberta. Since the glaciers of the last Ice Age melted away from Alberta, this river has eroded downward producing a deep, steep-walled valley that has a high incidence of landslides. Between the Alberta/British Columbia border and Fort Vermilion alone, at least 60 per cent of the valley walls have failed. This has severely affected development within the valley. The town of Peace River, in particular, has endured continuous slope instability problems affecting its housing, roads, and especially the Canadian National Railway. The Sagitawa Lookout on the Judah Hill Road provides an excellent view of both the townsite and the Peace River valley. Across the road from this lookout, you can see massive slumping along Highway 2 as it descends through the Heart River valley.

## THE STORY

There are many factors that contribute to the numerous earthflows and landslides that occur along the Peace River. Most of the valley walls consist of loose glacial and preglacial deposits of clay-rich sand and silt that sit on top of a weak shale bedrock. All of these deposits are easily and quickly eroded; and much of the valley through the town of Peace River is already 250 metres deep. At the bottom of the valley, the layer of sandstone upon which the bridge is built contains thin layers of unstable, crumbling shale that is prone to slope failure. As well, the abundance of the slippery impermeable clay throughout the sediments of the valley walls creates a situation of slow-draining soils and high water tables. As high soil moisture is critical in the evolution of landslides, the walls of the Peace River valley are constantly moving downslope.

The town of Peace River has suffered severe landslide problems since the 1950s when expansion began along the east wall of the valley. In particular, the railway has experienced difficulties, and track relocation and stabilization have been major challenges. The main problem area is near the junction of the Heart and Peace rivers. The slumping that began on Judah Hill in June 1984 realigned and damaged the track. Since then, remedial measures have been attempted such as construction of a timber retaining wall and toe berms, improved surface drainage of the Judah Hill Road, and removal of part of the crest of the hill. So far, railway traffic continues through this area, although more movement of the valley walls is expected. The cost of relocation and stabilization of this railway makes it the most expensive section of track in the entire province!

The Peace River is part of an ongoing study by geologists and engineers of slope stability problems along Alberta's river valleys.

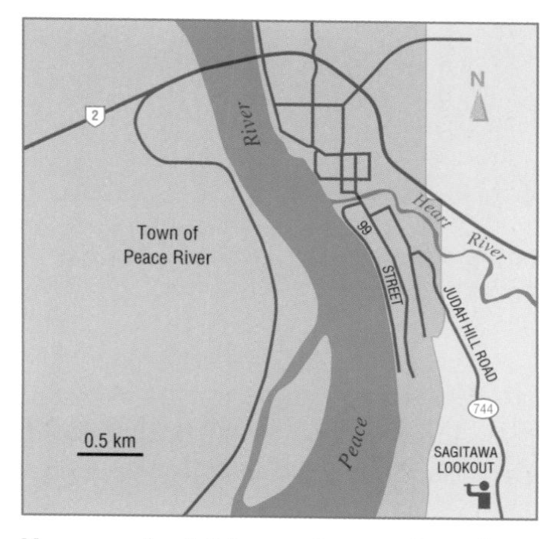


*Numerous landslides can be seen from the Sagitawa Lookout.*

*Milt Wright – British Columbia Archaeology Branch*

*Forest dropped by recent landslide near Rycroft, southwest of the town of Peace River. This landslide occurred in the same weak shale as those at Peace River.*

# BURNING SULPHUR NATURAL AREA: An Active "Bocanne"

*Alberta Environmental Protection*
*Steam and smoke smelling of sulphur rise for shale bedrock on the east slopes of the Buffalo Head Hills.*

## HIGHLIGHTS

On the east slopes of the Buffalo Head Hills, about 90 kilometres south of Fort Vermilion, bitter-smelling steam and smoke billows from rock debris that has slumped down the hillside. The steam is formed as the underlying bedrock of iron sulphide-rich marine shales oxidizes and produces great heat. Although not a common phenomenon, evidence of this process can also be seen along the Smoky and Peace rivers as well as in the "Smoking Hills" of the Arctic coast.

## THE STORY

This site is located on the Muddy River about three kilometres upstream from where it joins the Wabasca River in northern Alberta. The local bedrock is Cretaceous marine shale which normally is impermeable, but slumping along the river banks continues to produce a number of fractures in the shale. Once exposed to air along these fractures, the iron-sulphide minerals, such as pyrite, quickly oxidize and create temperatures of over 300°C. This heat bakes the nearby shale to a bright brick-red. Any groundwater that trickles down through fractures is quickly heated and turned into steam. Around the vents of escaping steam, native sulphur seeps through the fractured shale to solidify on the surface. Some of these sulphur seeps are currently burning and produce a very strong sour smell — a regular witch's brew!

In cooler weather, the smoke and steam lie in the river valleys as a heavy haze which early French Canadian fur traders called "boucanne" meaning clouds of smoke. The 19th-century geologist Robert Selwyn adopted the term and narrowed its meaning to naturally burning shale and altered its spelling to "bocanne." Likely, the Smoky River owes its name to bocannes that are still active along its valley.

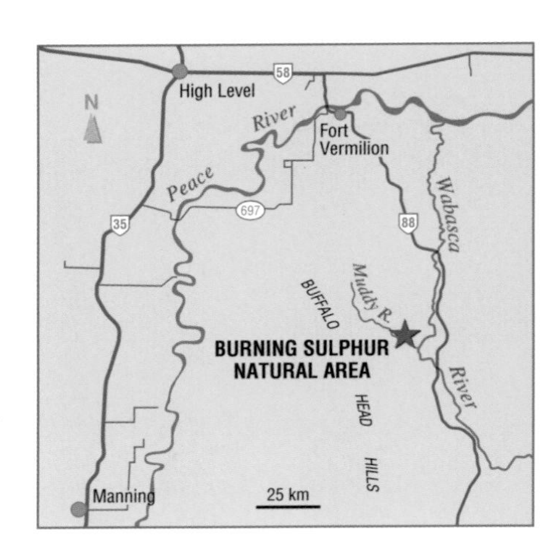

# HOT POT: A Burning Natural Gas Seep

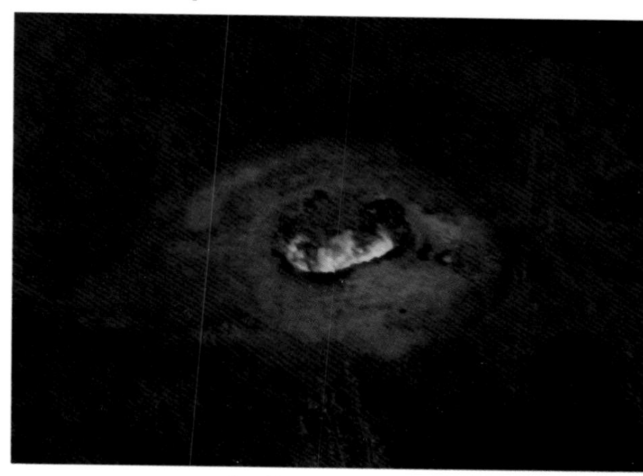

Ron Mussieux – Provincial Museum of Alberta

*Hot Pot north of High Level. Natural gas burns as it escapes from bedrock fractures.*

## HIGHLIGHTS

A rare geological phenomenon called the "Hot Pot" can be found 80 kilometres north of High Level, near Lutose Creek. Here, natural gas escaping from the earth's surface burns as a strange circular flare in the middle of the forest. The Hot Pot has been nominated as a Natural Area by the Alberta Land and Forest Service.

## THE STORY

Aboriginal Elders in the area talk of the "Hot Pot" as if it has existed forever. Their word for it is "kudadekune" which translates to English as "burning fire."

The fire burns all year long unless there is an exceptionally large snowfall; however, it is not long before it is relit by Aboriginal People. The flames shoot three to seven metres high and a barren flare pit encircles the fire. When it is not burning, the escaping gas causes the mud in the pit to bubble and churn.

What causes this odd fire? The answer is still unknown but numerous explanations have been suggested. One theory is that decaying subsurface organic matter is producing methane gas which escapes to the surface. The other, and more likely theory, is that this is a natural gas seep and because of the large amount of gas escaping, it is probably coming from a natural gas reservoir. The local bedrock is Cretaceous shale, normally an impermeable caprock in gas fields, but here it is highly fractured. Likely, the gas seeps into these fractures and makes its way up to the surface where it burns to create the Hot Pot. The Hot Pot is also located near the subsurface extension of the Great Slave Lake Fault. Could this ancient fault be involved in providing a route for the natural gas to escape? At this time, the answer is not known.

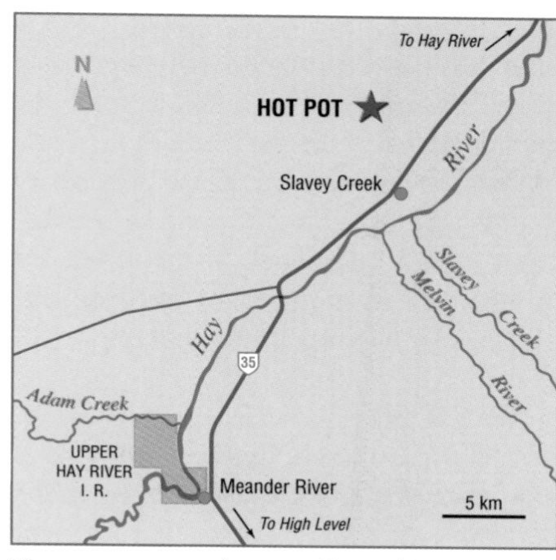


*There are no roads or trails to this site.*

# THE VERMILION CHUTES OF THE PEACE RIVER

A.2647 – Provincial Archives of Alberta

*The RCMP boat "Louise" being lowered over the Vermilion Falls.*

## HIGHLIGHTS

The Vermilion Chutes are about 80 kilometres downstream of Fort Vermilion. Canoeists define a chute as a short, steep, and narrow gap in a river channel where the water flows dangerously fast. These Chutes have been a major transportation barrier for everyone travelling on the Peace River since the earliest days of the fur trade in northern Alberta. The Chutes consist of a set of rapids and falls which extend over a distance of 2.5 kilometres — the rapids are impassable at high water level and the falls are only passable by portage.

## THE STORY

The Vermilion Chutes, named after the dusky red rock that outcrops there, are the result of changes in rock type along the Peace River's channel. After 80 kilometres of relatively quiet water from Fort Vermilion, huge boulders suddenly stretch across the river's path, signalling the beginning of the rapids. The boulders have been eroded from a 370-million year old layer of crumbly, grey dolostone.

The falls, a short distance downstream from the chutes, have formed where the river flows over a series of limestone ledges. The limestone is green with dark red mottling and unlike the dolostone contains many clay bands. The ledges in the riverbed were created because the limestone broke up along the clay bands and eroded away. At low water, the falls have a maximum vertical drop of four to five metres.

The Chutes are the only major obstacle to transportation along the Peace River. In the steamboat era, two ships

B. 3015 – Provincial Archives of Alberta

*Ledge of horizontal limestone forming the falls at Vermilion Chutes.*

operated on the Peace River; one upstream of Vermilion Chutes and the other downstream. Travellers and freight had to be removed from, and transported along a portage south of the Chutes.

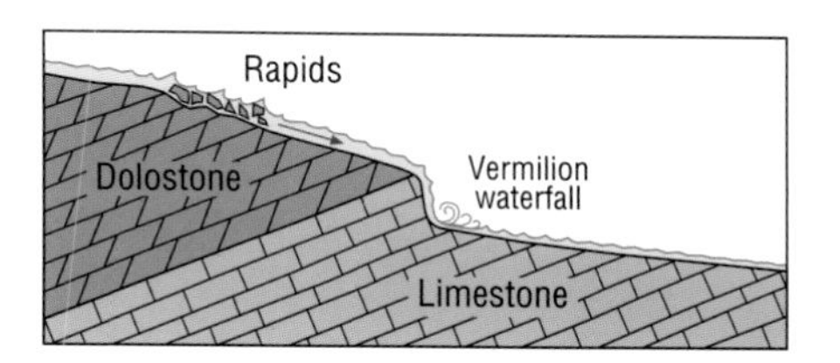


Today, Vermilion Chutes pose a problem only for the canoeing or kayaking enthusiast. If the water level is high, the rapids are classed as extremely difficult to almost impassable with a definite risk to life. All open canoes must portage. An eight-kilometre portage can be taken to go around the rapids and falls. The falls should never be attempted and they can be avoided by taking a short portage. Guidebooks are available for more information about trails and safe canoeing on the Peace River. The RCMP stress the importance of filling out a "travel registration form" if you are preparing to go into this wilderness area.

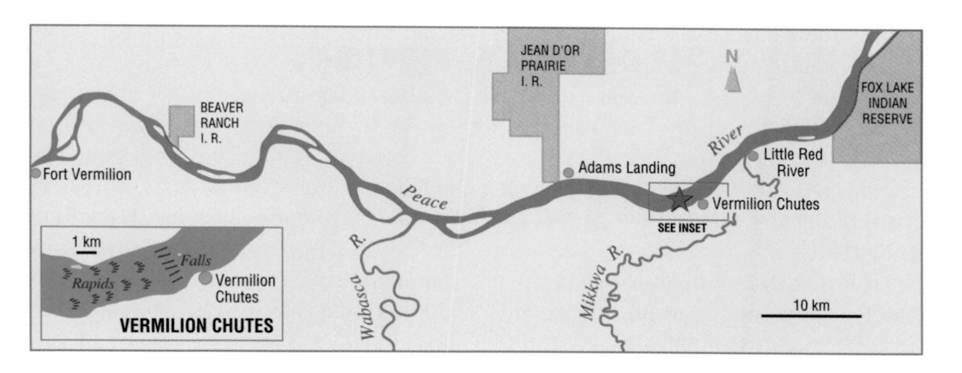

# THE SINKING LAND: Karst Landforms in Wood Buffalo National Park

Richard Stein – Alberta Geological Survey

*Near the Neon Lake spring are numerous small lakes with pink-tinted marl deposits. The milky white color of Neon Lake results from elemental sulphur and colonies of sulphur bacteria in suspension in the spring water.*

## HIGHLIGHTS

Wood Buffalo National Park contains the most extensive and best developed gypsum karst topography in North America. Karst is produced when groundwater dissolves soft rocks, such as gypsum, salt, and limestone, to form sinkholes, caves, underground streams, and sunken valleys. This is a fascinating environment because changes in the ground surface can occur very rapidly, occasionally within hours. Although many of the features are scattered throughout the park, the karst sites selected below are located near main roads and trails. Also in the park is the Salt Plains, in which large mounds of salt form where spring water, saturated with salt, bubbles up from the ground. The Salt Plains is a unique landform in Canada and is one of the natural features that led to the designation of Wood Buffalo National Park as a UNESCO World Heritage Site. In the less accessible western end of the park are spectacular spring deposits which feed both large and small lakes. Minerals precipitating from the springs have colored the lake bottoms and tufa deposits a delicate pink.

## THE STORY

Within Wood Buffalo National Park, the vast plain between the Caribou Mountains and the Slave River is underlain by marine sedimentary rocks of Middle Devonian age, about 390 million years old. These nearly flat-lying rocks contain tough, fractured limestones that alternate with easily dissolved gypsum layers that can reach thicknesses of 100 metres. Rain that falls on the eastern

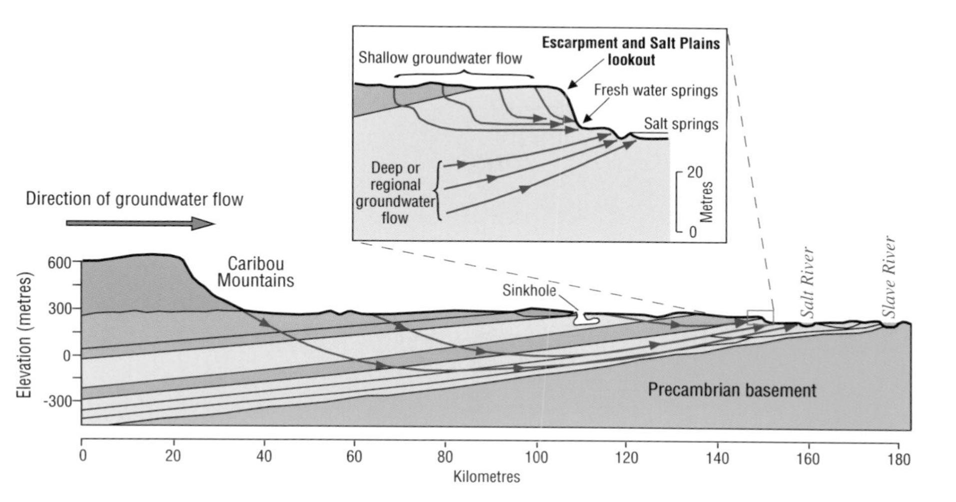


side of the Caribou Mountains and on the plain itself quickly trickles down through fractured rock to become subsurface groundwater. Even though limestone can be dissolved by this groundwater, the gypsum and salt are much more soluble. Thus, as the water flows through the gypsum and dissolves it, the overlying limestone layer acts as a protective caprock. When enough of the gypsum is dissolved, however, the limestone roof loses its support and collapses forming a variety of karst landforms that can be seen throughout this park.

*M. Peterson – Wood Buffalo National Park*

*Collapsing limestone in karst areas along the Rainbow Lakes Trail, Wood Buffalo National Park.*

# WOOD BUFFALO NATIONAL PARK

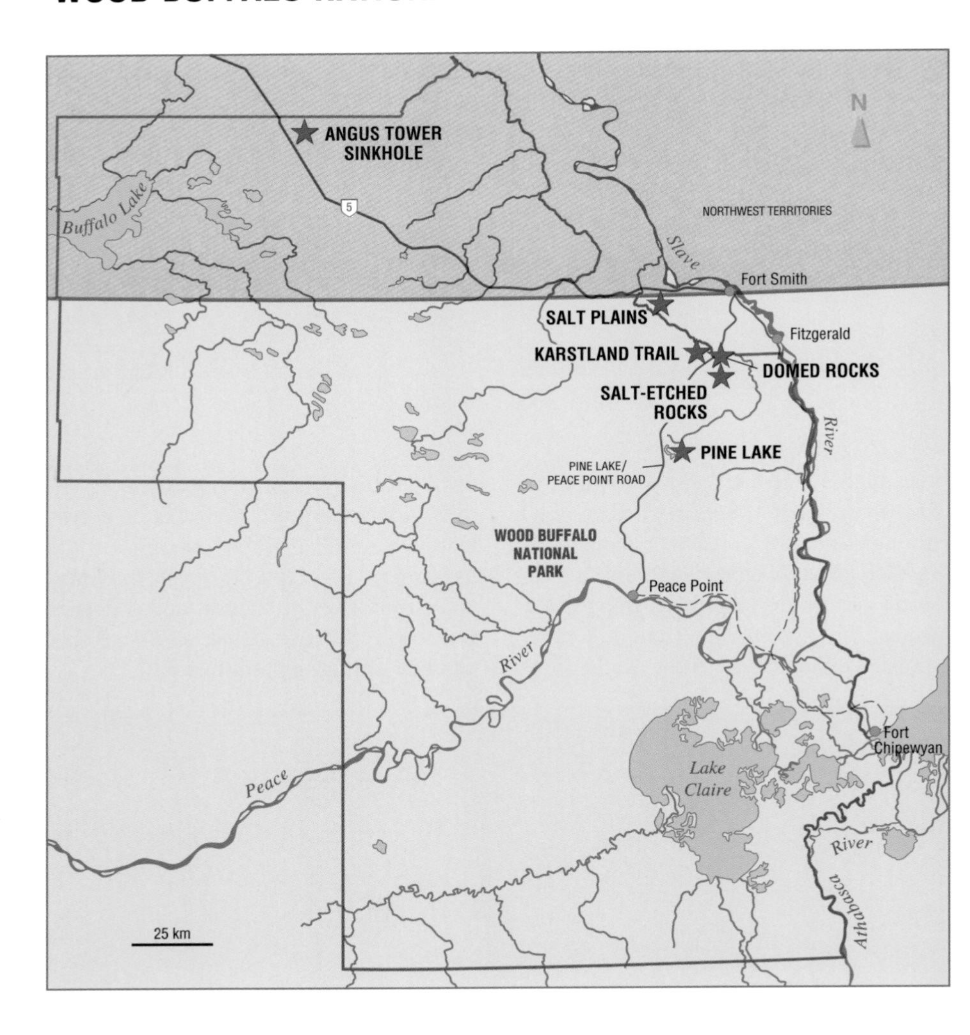

# ANGUS TOWER COLLAPSE SINKHOLE

*Wood Buffalo National Park*

*Group of sinkhole lakes.*

*Ron Mussieux – Provincial Museum of Alberta*

*This 60-metre deep sinkhole is the largest in the park.*

The most abundant karst features in the park are collapse sinkholes such as the one at the Angus Fire Tower. Here, the roof of an underground cavern collapsed and formed a circular depression, 60 metres deep and 100 metres across, making it the largest sinkhole (or doline) in the park. Sinkholes can occur singly or in groups — an area near the Rainbow Lakes Trail boasts 50 sinkholes in a four square kilometre area. Caves are often associated with collapse sinkholes because the flat gypsum layers are dissolved horizontally beneath the unstable limestone. These caves often contain permafrost and sometimes detached blocks are supported by thin films of ice. Caves are highly dangerous and entering them is prohibited. Check with Park personnel before exploring any of these sites.

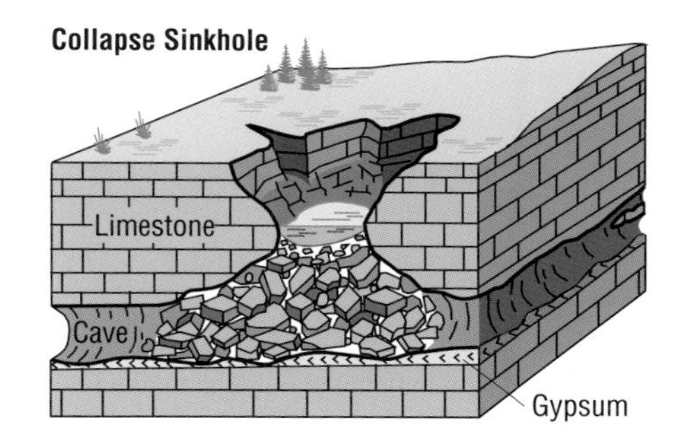

# SALT PLAINS LOOKOUT

*McCloskey – Wood Buffalo National Park*

*An aerial view of the Salt Plains. White patches are salt flats, not patches of snow.*

The Salt Plains covers an area of 370 square kilometres in Wood Buffalo National Park. On the west side of the plains is an escarpment that rises 60 metres high and is made up of layers of dolostone, limestone, gypsum, anhydrite and rock-salt that were deposited in a shallow, salt-saturated sea during the Devonian Period, about 390 million years ago. From the base of this escarpment emerge two kinds of springs: fresh water springs found close to the base and salt water springs found about five metres further east. The fresh water springs have a heavy flow and are formed by groundwater that has seeped down through the escarpment. The salt springs, on the other hand, are formed from groundwater that has seeped eastwards from as far away as the Caribou Mountains. This water has dissolved salt beds along the way and eventually emerges as salty, bubbling springs. Some of these springs carry ten times more salt than the equivalent amount of sea water! The brine coats the surrounding clays with a salty slime and, as the water evaporates, cubic salt crystals grow and form shimmering white salt flats across the plain. During dry years, the mounds of salt around a spring can reach a metre in thickness. The brine springs flow eastwards to form the Salt River which slowly meanders its way across the salty mud flats. In some places, the river is only three metres wide and one metre deep!

Historically, the use of these salt deposits constitutes one of Alberta's oldest industries. For hundreds of years, Aboriginal People used the salt to cure fish and meat and to tan hides. During the 1800s, both the Oblate missionaries and the Hudson's Bay Company harvested the salt and shipped it throughout the north. As late as 1920, the Hudson's Bay Company was mining five tons of salt annually and selling it at Fort Smith at the princely sum of 10 cents a pound.

*Richard Stein – Alberta Geological Survey*

*Cubic salt crystals (halite) growing in a saline pool.*

The salt attracts many mammals and birds, and their tracks can often be seen in the barren salty mud. A panoramic view of the Plains can be seen from Salt Plains Outlook, 35 kilometres west of Fort Smith and 16 kilometres south on Parson's Lake Road, which is a summer road. A short trail leads down to the Plains themselves.

*Wood Buffalo National Park*

*During periods of low rainfall and high evaporation, salt will build up into mounds around the saline springs.*

# KARSTLAND TRAIL AND ITS "DRUNKEN FORESTS"

*Ron Mussieux – Provincial Museum of Alberta*
*Wooden walkway with interpretive signs describing gypsum outcrop, Karstland Trail.*

*Ron Mussieux – Provincial Museum of Alberta*
*Tilting trees along Karstland Trail are locally called "drunken forests."*

Karstland Trail is a well developed, one-kilometre interpreted hike which starts from the parking lot of the Salt River picnic site and passes through a heavily wooded area of collapse sinkholes. These sinkholes form when groundwater dissolves the gypsum below the ground surface until caverns and connected passages are formed. Eventually, the limestone roof will collapse onto the cavern floor, and a sinkhole is created. The sinkholes along Karstland Trail are often hidden by spruce-poplar forests but can be found by looking for tilting or fallen trees. These are locally called "drunken forests."

This gypsum karst is geologically young and has only been forming since the melting of the last continental glaciers, some 10,000 years ago. It is a very active and sometimes dangerous area as new sinkholes are constantly forming. Occasionally, trails collapse and must be closed temporarily.

This area is of particular biological interest because it contains underground cavities below the frost line. Here, red-sided garter snakes gather and hibernate to survive the long, cold winter. These snakes are North America's most northerly reptiles.

# ROCKS THAT SWELL: Domed Rocks at the South Salt River Bridge

*Don Taylor – Provincial Museum of Alberta*
*Rock layers are domed by an increase in volume as anhydrite alters to gypsum.*

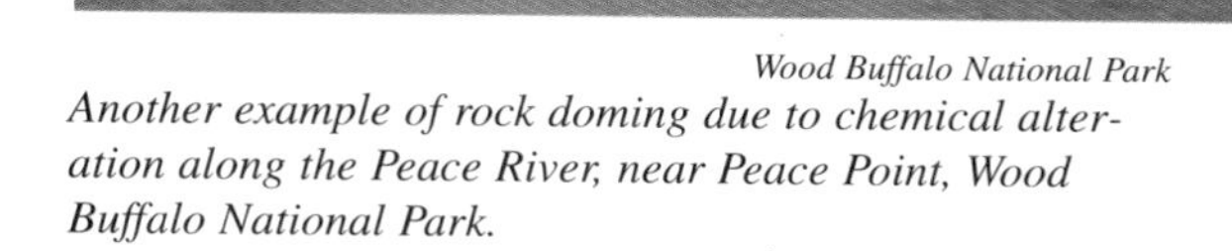

*Wood Buffalo National Park*
*Another example of rock doming due to chemical alteration along the Peace River, near Peace Point, Wood Buffalo National Park.*

East of the picnic area, near the South Salt River bridge, is a steep outcrop of gypsum that is domed upwards into a structure that geologists call an anticline. Unlike anticlines in the Rockies, this structure was not formed through mountain building but by a chemical process. Much of this gypsum was originally deposited as the mineral anhydrite (calcium sulphate). As erosion removed the rock above it, water was able to enter into the unstable anhydrite and convert it into gypsum (hydrated calcium sulphate). This alteration is accompanied by an increase in volume of 30-50 per cent and the swelling buckled the overlying rock layer and forced it upwards into an arch, or anticline. A short hike north of the picnic grounds along the north loop of the Salt River trail will bring you to a large depression that is described as a "collapsed dome."

# THE PECULIAR SALT-ETCHED ROCKS OF GROSBEAK LAKE

*Ron Mussieux – Provincial Museum of Alberta*

*A granite boulder is severely weathered by the growth of salt crystals in cracks or holes on the rock's surface.*

The mud flats surrounding Grosbeak Lake have been described by Park staff and visitors as a bleak, barren moonscape. The hundreds of boulders that lie scattered across the red clay flat have been sculpted into a variety of odd shapes by salt weathering. The salt comes from nearby brine springs. Weathering is the disintegration of rocks and minerals on the earth's surface by either physical processes (such as rain, wind, heat, frost) or by chemical and biological processes.

The Grosbeak Lake boulders are erratics that were transported here by glaciers over 10,000 years ago. Most are granites, gneisses, and sandstones that came from the Canadian Shield to the northeast and are normally extremely hard and durable. At this location, however, these tough boulders are attacked and ultimately destroyed by an unusual type of weathering that also occurs in the deserts of Iran, Australia, Chile, and the western United States.

Salty springs that trickle from the base of the escarpment south and west of Grosbeak Lake supply the water that meanders across the boulder-strewn mud flat towards the lake shore. This water contains ten times as much salt as normal sea water! A continuous supply of this water covers the lower surface of the boulders and penetrates cracks, rock pores or fractures. During dry periods, salt crystals begin to grow in the rock cavities as the spring water evaporates. As the crystals grow, they exert pressure on the surrounding rock walls. They eventually wedge apart and force out rock fragments, slowly creating the peculiar boulder landscape we see. Porous rocks are disintegrated grain by grain, nonporous rocks are simply split, while other rocks are notched on their lower surfaces. Occasionally a rock is attacked from all sides and holes are cut right through it. The entire process is further enhanced by frost fracturing in the boulders during the spring and fall.

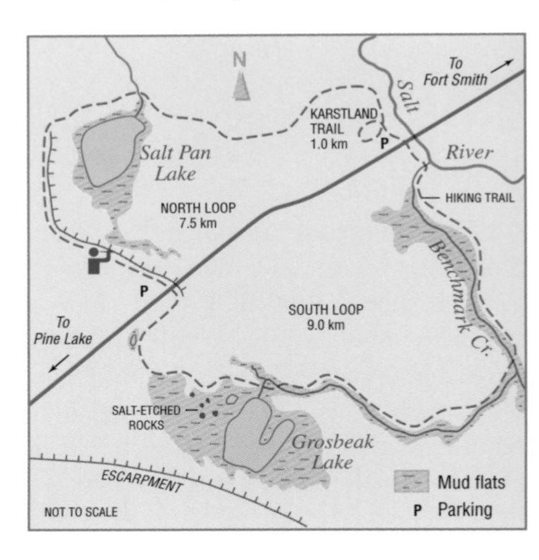

# PINE LAKE: A Large Sinkhole Lake

*Wood Buffalo National Park*

*The variation of water color in Pine Lake reflects the outlines of individual sinkholes which have joined together to form the lake.*

The basins of Pine and Rainbow lakes are solution sinkholes. These form where water pools in a shallow depression on the ground surface and trickes into fractures and cracks dissolving underlying sediments. When these cracks are large enough, the loose surface sediments are then washed down, plugging the underground drainage. A shallow crater is created which often fills with water from rainfall and nearby streams, forming small lakes. Some are temporary, and there are examples of small lakes that have drained overnight, much to the surprise of park visitors! Pine Lake is actually composed of five connected sinkholes, and near the picnic grounds the variation in water color highlights the underwater rim between two of the sinkholes. Pine Lake is likely part of a larger underground drainage system. Looking at a map of Alberta you can see several other sinkhole lakes that extend in an almost straight line southeast of Pine Lake for about 15 kilometres.

# THE MOUNTAIN RAPIDS OF THE SLAVE RIVER

*Natural Resources Service*

*An aerial view of the granite peninsula and the Mountain Rapids. A trail down the middle of the peninsula follows the power line and leads to the rapids.*

## HIGHLIGHTS

The Mountain Rapids are one of a group of four rapids, collectively called the Slave River Rapids. This spectacular section of the Slave River is one of the finest examples of rapids in North America and has been designated as a National Historic Site because of its significance as a transportation barrier that affected the development of Canada's north. Today, these violent and dangerous rapids have become a tourist destination for expert kayakers and river rafters.

## THE STORY

For most of its course, the Slave River follows the boundary between the igneous and metamorphic rocks of the Cambrian Shield to the east and the sedimentary rocks of the Interior Plains to the west. The Slave River discharges over 80 per cent of all of Alberta's combined river flow. June's average monthly discharge at Fitzgerald is a remarkable 6000 cubic metres per second, which is about 11.5 times the flow of the North Saskatchewan River at Edmonton. Between Fitzgerald and Fort Smith, a distance of 27 kilometres, the Slave River drops 33.2 metres over four sets of rapids: the Cassette, Pelican, and Mountain rapids, and the Rapids of the Drowned. All four of these rapids are formed by granite ridges of the Canadian Shield which cross the river channel in a northeast/southwest direction. These ridges are the only easily accessible outcrop of Canadian Shield west of the Slave River.

The Mountain Rapids are immediately east of a long granite peninsula which cuts the river channel in half. Before the construction of the portage road on the west side of the river, the other three sets of rapids were bypassed on the east using narrow side-channels. The Mountain Rapids, however, could not be passed on the east, so the Mountain Portage was established at the narrowest neck of the peninsula. Scows were winched to the summit and allowed to slide down the other side until they reached the river.

Several trails parallel the powerline and lead to the end of the peninsula. The roar of turbulent water can be heard long before you reach the viewpoint that overlooks the rapids. The pink granite islands in the channel are an extension of the peninsula which has been breached in several areas by the river. The islands just below the first rapids are the world's most northerly nesting area of American White Pelicans (*Pelecanus erythrorhynchos*).

Most of the granite exposed at the rapids has been smoothed, polished, and striated, not by the river but by a southward advancing continental glacier over 10,000 years ago. These elongated rounded glacial landforms are called "whalebacks" and their shape indicates the direction of ice flow. The low, rounded ends point in the direction from which the ice came, while the down-ice end is rough and angular where the glacier plucked off blocks of rock as it passed over.

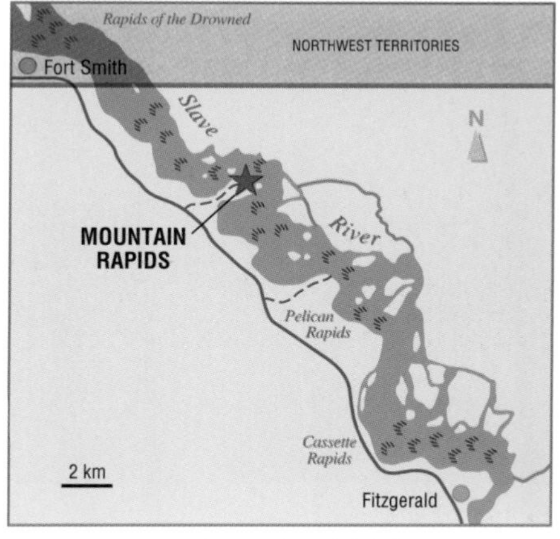


*Mountain Rapids Trail is 6.3 kilometres south of Fort Smith town limits.*

*B.2880 – Provincial Archives of Alberta*

*A fur-trading scow runs a portion of the more navigable Slave River Rapids near the Mountain Rapids. At Mountain Rapids the boats had to be portaged over the peninsula.*

# PELICAN RAPIDS: An Outcrop of Canadian Shield

*Tom Chacko – University of Alberta*

*Aerial view of Precambrian Shield outcrop at Pelican Rapids.*

## HIGHLIGHTS

The granite ridge forming the Pelican Rapids is one of the most accessible outcrops of Canadian Shield west of the Slave River. The Canadian Shield is the core of the North American continent and contains rocks nearly four billion years old, the oldest known in the world. These ancient Precambrian rocks of the Shield represent over 85 per cent of our planet's entire geological history. The granite of Pelican Rapids tells a story of collisions between continents two billion years ago.

## THE STORY

The Canadian Shield is exposed at the surface over about 50 per cent of Canada and it underlies an even larger portion of the North American continent. To many Canadians, the Shield is a frontier — a vast, empty, inaccessible land of barren rock, muskeg, lakes and, in the south, impenetrable forests. Although there is little fertile soil, the Shield does hold much of Canada's mineral wealth including gold, silver, uranium, iron, nickel, copper, and even diamonds.

The Shield is actually not a single entity but rather a combination of several small continental fragments (or protocontinents), each with its own geological history. Collisions between these proto-continents welded them together and resulted in the formation of ancient mountain ranges; these have endured millions of years of erosion by water and glacial ice until finally their crustal roots were exposed. These roots, then, are the Shield that we see today.

Between 2.0 and 1.9 billion years ago, the broad belt of igneous rock in northern Alberta, called the Slave Granite, was formed by the collision of a small continental fragment with a larger proto-continent. Metamorphosed sedimentary rocks were caught up in this collision and at depths between 18 and 24 kilometres into the crust, they began to melt and form a granitic magma. This magma eventually crystallized at 900°-1000°C and formed the Slave Granite which still contained surviving blocks of the metamorphosed sedimentary rocks from which it had formed.

At Pelican Rapids, the glacially scoured peninsula projecting into the river gives the visitor a window to view the results of these processes which occurred deep within the earth. You can see large blocks of metamorphosed rocks, called gneiss, that appear to float within the granite. These dark-colored gneisses were originally shales, and the light-colored blocks were once sandstones. Surprisingly, despite having been buried deep in the crust and surviving a continental collision and partial melting, the original sedimentary bedding in the sandstones can still be seen.

The granite at Pelican Rapids contains several high-temperature minerals such as garnet, cordierite, and hercynite, proving that it formed at 900°-1000°C, rather than the usual 700°C for typical granites.

Tom Chacko – University of Alberta

*Large blocks of metamorphosed sedimentary rocks are contained within the granite. The dark-colored gneiss was once shale. The light-colored sandstone shows original bedding structure.*

Tom Chacko – University of Alberta

*Large mineral crystals within the almost two-billion-year old Slave Granite.*

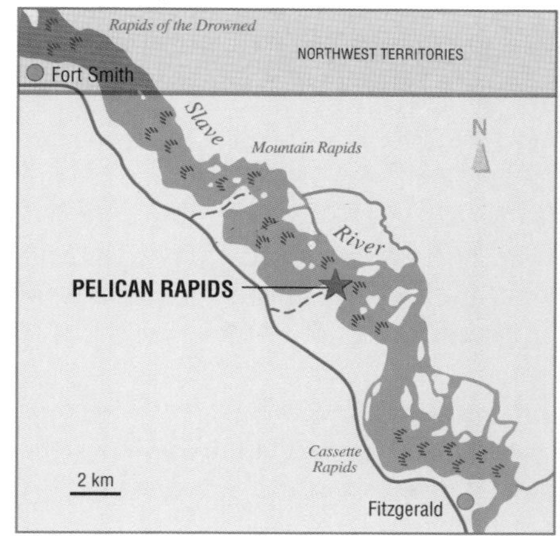


*Hiking trail is 10.3 kilometres south of Fort Smith.*

# PEACE-ATHABASCA DELTA: The World's Largest Inland Freshwater Delta

C. Wallis – Natural Resources Service
*Sediment laden channels deposit their load into shallow lakes and ponds slowly adding to the Peace-Athabasca Delta.*

Wood Buffalo National Park
*Trees are confined to the better drained banks of the channels in the Peace-Athabasca Delta.*

## HIGHLIGHTS

The Peace-Athabasca Delta is a vast maze of sand bars, ponds, sedge meadows, and meandering channels covering almost 4000 square kilometres. With the evolution of this diverse landscape, a unique ecosystem has developed, making this delta an irreplaceable resource. Reasonably priced air tours of the Peace-Athabasca Delta can be arranged from Fort Smith or Fort Chipewyan and boat tours with licensed guides can be arranged from Fort Chipewyan.

## THE STORY

A delta is an accumulation of sediments at the mouth of a river where it enters a larger body of water. When a river flows into a lake or sea, its speed decreases and the sediments carried along in its current sink to the bottom. These sediments form a low-lying, often marshy delta that slowly spreads farther out from shore. The Peace-Athabasca Delta is formed by two converging deltas: the Peace River Delta, which is no longer actively growing, and the Athabasca River Delta, which is spreading eastwards across the west end of Lake Athabasca.

A peculiar annual cycle exists here that produces a summer flooding of the Delta. Flooding is vital to its environment and has fostered an ecosystem dependent upon fluctuating water levels. For most of the year, the level of the Peace River is below that of Lake Athabasca and the lake water flows unimpeded northwards via Rivière des Rochers and into the Peace. The joining

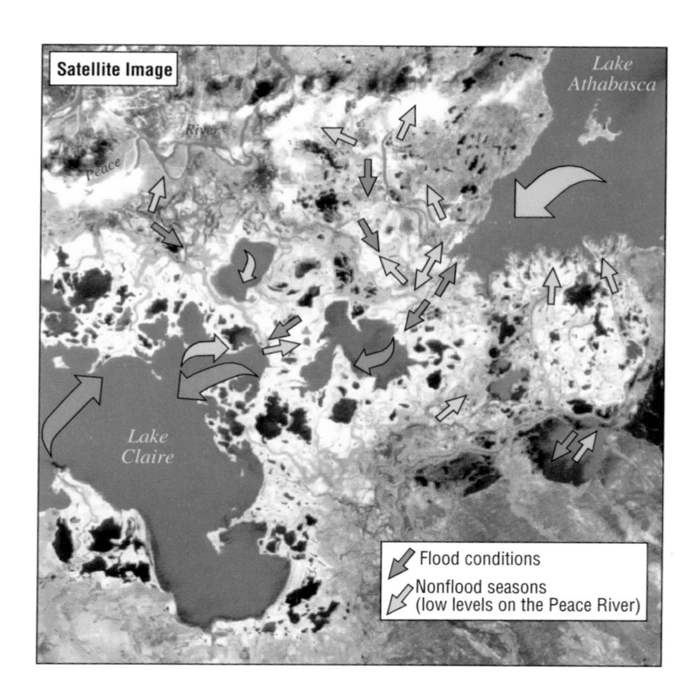


of these two rivers creates the Slave River. During this time, delta waters flow into Lake Athabasca and the delta is drained. In early summer, however, the Peace River rises higher than Lake Athabasca and the flow of the lake into the Slave River reverses. The lake backs up and spills out to flood and replenish the Delta. When the Peace River returns to normal levels, Waters from Lake Athabasca once again flow north and the delta waters recede. Thus, the Peace River is the key factor in the summer flooding. This age-long cycle was broken in the 1970s by the construction of the W.A.C. Bennett Dam on the Peace River in British Columbia. When the Peace was dammed, Lake Athabasca drained year-round and the flood cycle ended. The drying of the Delta became the focus of public controversy and, since then, management of water levels in the dam has returned the Delta to its natural state.

The Peace-Athabasca Delta began to form near the end of the last Ice Age as glaciers melted away from the west end of Glacial Lake Tyrrell. Since then, the exposed glacial lake sediments have been carried by the Peace and Athabasca rivers into Lake Athabasca and deposited on the evergrowing Delta. The exposed sand of the delta is also being blown into dunes along the south shore of the lake, forming the largest active dune field in Canada.

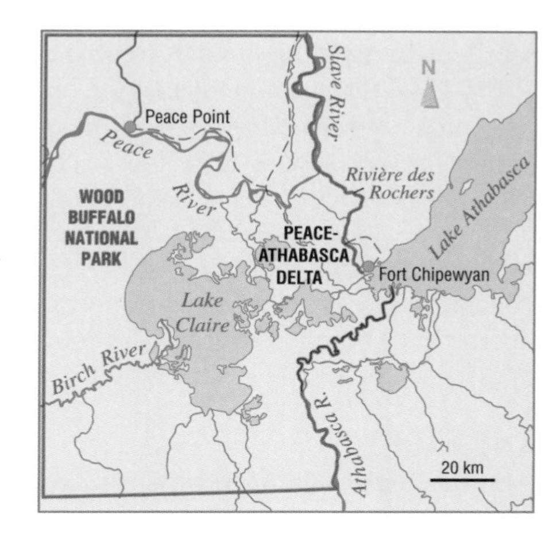

# THE ATHABASCA DUNES ECOLOGICAL RESERVE:
## Alberta's Largest Active Dune Field

*Pat McIsaac – Natural Resources Service*

*Aerial view of Athabasca sand dunes encroaching on forests and filling the lakes and rivers.*

## HIGHLIGHTS

Many Albertans are unaware that northern Alberta has actively migrating sand dunes. The Athabasca Dunes Ecological Reserve, together with a small dune field west of Richardson River, is the largest area of moving, or migrating sand in Alberta, as well as part of the largest dune field in Canada. This area is an excellent site to study how dunes form. Because of their remoteness and enormous size, the Athabasca Dunes are best viewed by air as you travel from Edmonton or Fort McMurray towards Fort Chipewyan.

## THE STORY

In this dune field, massive quantities of sand and strong winds have created a environment of blowing, shifting sand. The dunes bury and destroy jack pine forests and fill in lakes and marshes, leaving behind a flat, barren plain. In contrast, the surrounding area presents forested hills, lakes, and marshes and is thought by many to be one of the most beautiful areas in the province.

The Athabasca dune field is seven kilometres long, 1.5 kilometres wide, and some of the dunes reach heights of 35 metres. The field consists of two enormous crescent-shaped dunes that have merged and migrated at least ten kilometres eastwards to their present location. On the dune field itself are thousands of small sinuous dunes that migrate at right angles to the northwest direction of the prevailing winds. The entire field travels southeastward at a rate of 1.5 metres per year, adding new sand to its bulk as it advances.

What sort of geological process provides an area with sand nearly 35 metres deep? The ultimate source is the granite, gneiss, and sandstone bedrock found in northeast Alberta. Only these rock types provide massive amounts of clean sand. During the last Ice Age, glaciers eroded this bedrock and ground it up into vast quantities of sand. As the glaciers melted, torrents of water poured into glacial lakes and deposited the sand. When the lakes eventually drained, the sand was dried by the winds and blown into large dunes.

This environment of blowing sand has existed for 8000 years. Any vegetation which begins to stabilize the dunes is destroyed by the wild fires so common in northern Alberta. Geologists speculate, however, that the migration of these dunes may finally stop when they reach the Maybelle River.

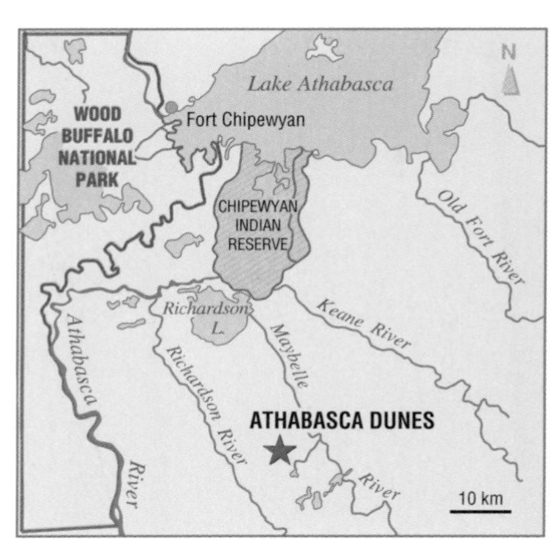

*Brett Purdy – University of Alberta*

*Wall of advancing sand burying a forest.*

# ATHABASCA OIL SANDS: Tapping Canada's Most Valuable Energy Resource

*Syncrude Canada Inc.*

*Aerial view of oil sands open pit mining at Syncrude's Mildred Lake plant, 40 kilometres north of Fort McMurray.*

## HIGHLIGHTS

Trapped in the Athabasca oil sands in northeast Alberta are nearly 1 trillion barrels of bitumen — more than the conventional reserves in the Middle East, United States, and western Europe combined. As the world's conventional oil reserves are depleted, these sands will become essential to supplying Canada's energy needs. Although they are the world's largest single oil sand deposit, less than 15 per cent of the bitumen can be extracted with current technology.

## THE STORY

Oil sands are a compacted sand containing heavy oil, or bitumen. About 12 per cent of the Athabasca oil sand is bitumen, a tar-like mixture of hydrocarbons that requires chemical alteration to make a lighter oil that is transportable and marketable. Numerous hypotheses have been suggested to explain the origin of the vast amounts of bitumen in these sands. Some geologists think that lighter oil originally formed elsewhere in older rocks, possibly the underlying Devonian limestone, and migrated upwards into the Cretaceous sands where it then dried. Others think the bitumen formed in the sand where it is now found.

Unlike conventional oil reserves, some of the Athabasca oil sands are near the surface and can be economically mined by open pit methods. Massive machinery removes the muskeg and forest-covered overburden and hauls the oil sands to the surface where it is conveyed to an extraction plant. Here, steam and hot water separate the sand from the bitumen, which is then diluted with naphtha, heated to 500°C, and upgraded to make the final product, called synthetic crude. Synthetic crude, so called because of its extensive refining process, is a light, high-quality, sweet crude oil that is then piped to refineries in Edmonton where it can be processed into gasoline, diesel, and other products. It takes about two tonnes of oil sand to produce one barrel (159 litres) of this synthetic oil!

The Athabasca oil sands is the only deposit of its type in the world to attract large-scale commercial development. Suncor Energy began production in 1967 and became the first successful oil sands venture worldwide. Syncrude Canada Inc., built in 1978, is the world's largest producer of synthetic crude, as well as one of the largest earth-moving operations, covering a 40,000 hectare area. Together, these two mines produce more than 280,000 barrels of oil per day, supplying nearly 20 per cent of Canada's petroleum needs.

The Oil Sands Interpretive Centre in Fort McMurray is open year-round, and in the summer, tours to Syncrude and Suncor depart from there. Information and reservations can be obtained at the visitors' bureau.

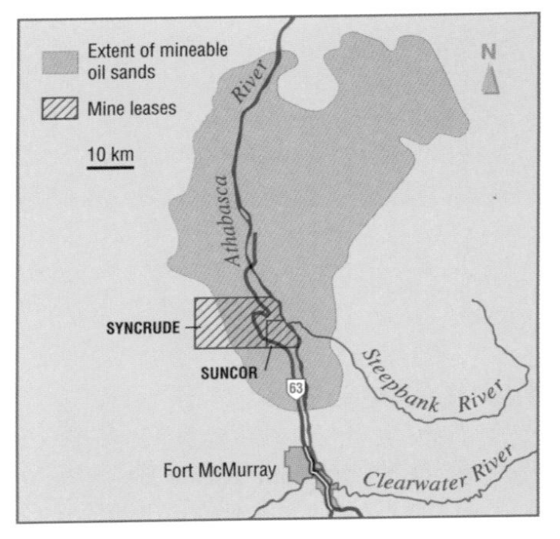

Syncrude Canada Inc.

*Oil sands being mined by shovel and 240-tonne truck, Mildred Lake plant.*

# ATHABASCA RIVER VALLEY AT FORT MCMURRAY

Don Taylor – Provincial Museum of Alberta

*One hundred-million-year old dark oil sands sit directly on top of 345-million-year old white limestone. Geologists call a gap in the rock record, such as this, an unconformity. Athabasca River at Fort McMurray.*

## HIGHLIGHTS

The city of Fort McMurray is located at the junction of the Athabasca and Clearwater rivers in northeast Alberta. The rocks exposed along the river valleys, and the shape of the valleys, tell a variety of geological stories about glaciation, a catastrophic flood, and one of the most conspicuous "unconformities" in the province. Fort McMurray is where Aboriginal Peoples first used the bitumen from Athabasca oil sands to caulk their canoes.

## THE STORY

The most interesting and visible geological feature of the river valley is the marked color contrast between the rock units exposed along the valley walls. Looking at the photo, you can see dark bitumen-impregnated sands lying on a white wavy surface of limestone. The sandstone dates from the Cretaceous Period, about 100 million years ago, while the limestone is from the Devonian Period, about 345 million years ago. Thus, between these two rock layers, there is a 245-million-year gap in the rock record! This gap, called an unconformity, reflects a long period of erosion and periods of no deposition of other sediments. One useful test of age above and below an unconformity is to look at the fossils present. Here, the limestone bed contains a wide variety of marine fossils, such as brachiopods, while the overlying sandstone contains much younger wood and clam fossils, from close to the shore on a delta.

At Fort McMurray, the valleys of both the Clearwater and lower Athabasca rivers appear much too wide for their rivers. This is particularly noticeable with the Clearwater River, which meanders back and forth within its valley. These misfit rivers, plus the extensive gravel and

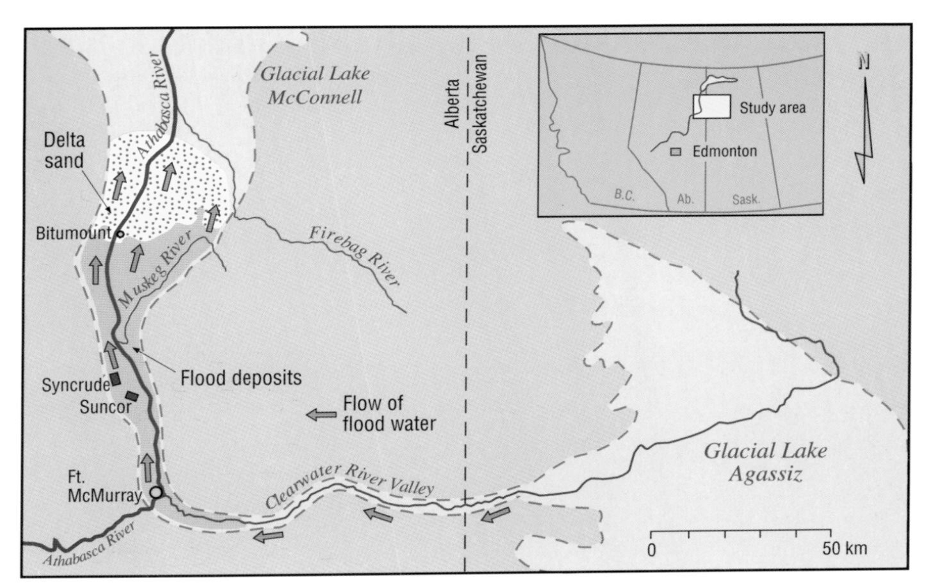


*Modified after D.G. Smith and T.G. Fisher (1993) Geology, V. 21, pg. 9.*

boulder deposits between Fort McMurray and Bitumount, have naturally led to geological debates. Recent investigations suggest that these features are the result of a megaflood that originated in an ancient lake called Glacial Lake Agassiz. As the continental glaciers melted, the lake filled and eventually burst over a drainage divide and poured westwards. The surging water quickly eroded and enlarged the Clearwater and lower Athabasca valleys and dumped its load of gravel near Fort McMurray; its sands and silts produced the evergrowing Glacial Lake McConnell delta. These sediments contributed to the present day Peace-Athabasca Delta. A radiocarbon date on fossil wood in these sediments indicates the flood occurred around 9900 years ago. Calculations suggest that it lasted for several years, and about 21,000 cubic kilometres of water drained from Lake Agassiz, lowering its water level by 52 metres. This water then flowed into the Arctic Ocean and likely raised global sea levels by six centimetres — not an insignificant amount of water!

The Fort McMurray area has made a significant contribution to Alberta's economy. Salt mining was once an important industry and, in fact, the wavy surface of the Devonian limestone along the river valley reflects collapsed salt beds below the surface. The major industry in the region today is the Athabasca oil sands.

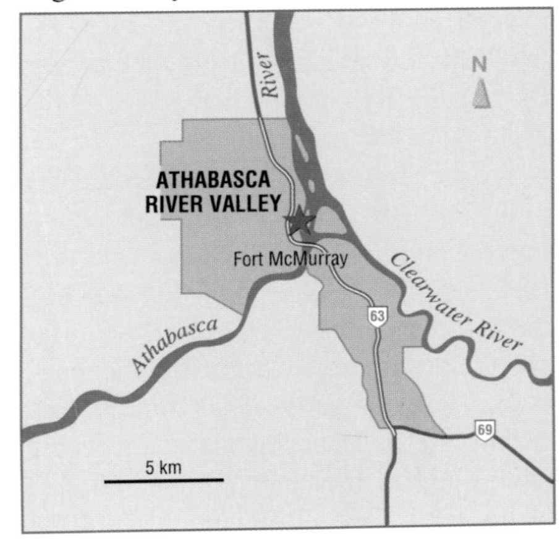

# THE HISTORICAL GRAND RAPIDS OF THE ATHABASCA RIVER

*John Kramers – Alberta Research Council*

*The spectacular and ever-dangerous Grand Rapids are produced by the rapidly flowing river spilling over large concretions exposed on the floor of the river bed.*

## HIGHLIGHTS

The Grand Rapids are on the Athabasca River between Athabasca and Fort McMurray. The rapids are formed by huge boulders that stretch across the river from bank to bank creating a major transportation hazard and a striking geological feature. During the fur trade, the Athabasca River became the Hudson's Bay Company's main river route to the north and the notorious Grand Rapids were the greatest obstacle on it. The rapids are still remote today and can be approached only by boat or aircraft. Because of the isolation of this area and the difficulty of the rapids, a canoe tour should be attempted only by the most experienced canoeists. The Grand Rapids are impassable at any water level and have claimed the lives of a number of unsuspecting and ill-prepared canoeists.

## THE STORY

Travelling the historic boat route down the Athabasca River from Athabasca (formerly called Athabasca Landing) to Grand Rapids, one reaches the tiny settlement of Pelican Portage, about 75 kilometres above the rapids. This was the site of a 19th-century gas well blowout that burned out of control for 21 years producing a bright, wild beacon to river traffic. The first sign of the Grand Rapids comes five kilometres upstream where you can begin to hear the roar of the rapids. At the rapids, 253 kilometres downstream of Athabasca Landing, a large island separates the river into an east channel and a more dangerous west channel. The rapids run for 1.6 kilometres dropping over 11 metres in the first kilometre over a sloping boulder dam.

On the island, little remains of the

tramway that the Hudson's Bay Company built at the turn of the century to help transport goods on the Athabasca River. Goods were transported downstream on flat-bottomed boats called scows. At the rapids, the cargo was unloaded onto the island and transported on the tramway while the boats were lined down the east channel.

The rapids are the result of river erosion of the 110-million-year old sandstone of the Grand Rapids Formation. This formation, which forms the large, nearly vertical outcrop on the east side of the valley, is divided into three major sandstone layers. The lowermost layer creates the rapids because it is filled with large, two-to-three-metre wide, concretions that often contain pieces of petrified logs. These concretions were formed in a similar fashion to those at Red Rock Coulee (see page 232). As the river erodes away the sand matrix, these huge concretions come loose and dam the river bed.

West of this region, in the Wabasca Lakes area, the Grand Rapids Formation is well known as a subsurface oil sand deposit. The oil is found in the upper two layers of the formation while there is no oil in the lower boulder-rich layer that is responsible for the rapids.

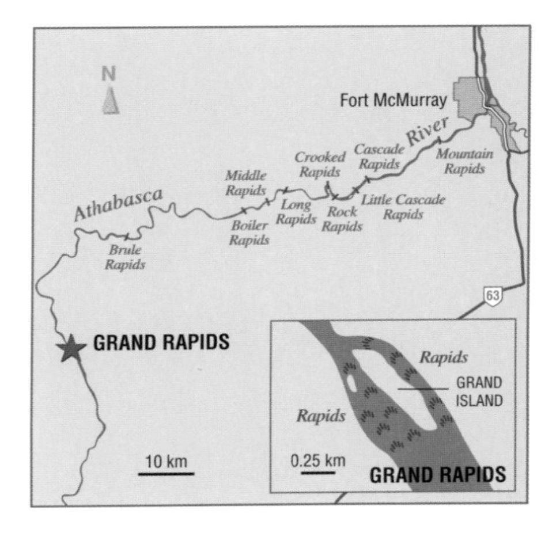

B. 2890 – Provincial Archives of Alberta

*A flurry of activity at the Grand Rapids as Hudson's Bay Company fur boats arrive. The cargo is unloaded before the boats are dragged through the rapids.*

# CHAPTER 3: EDMONTON AND AREA

*Aerial view of prairie "doughnuts," or prairie mounds, south of Edmonton.*
*Dixon Edwards – Alberta Geological Survey*

# EDMONTON AND AREA

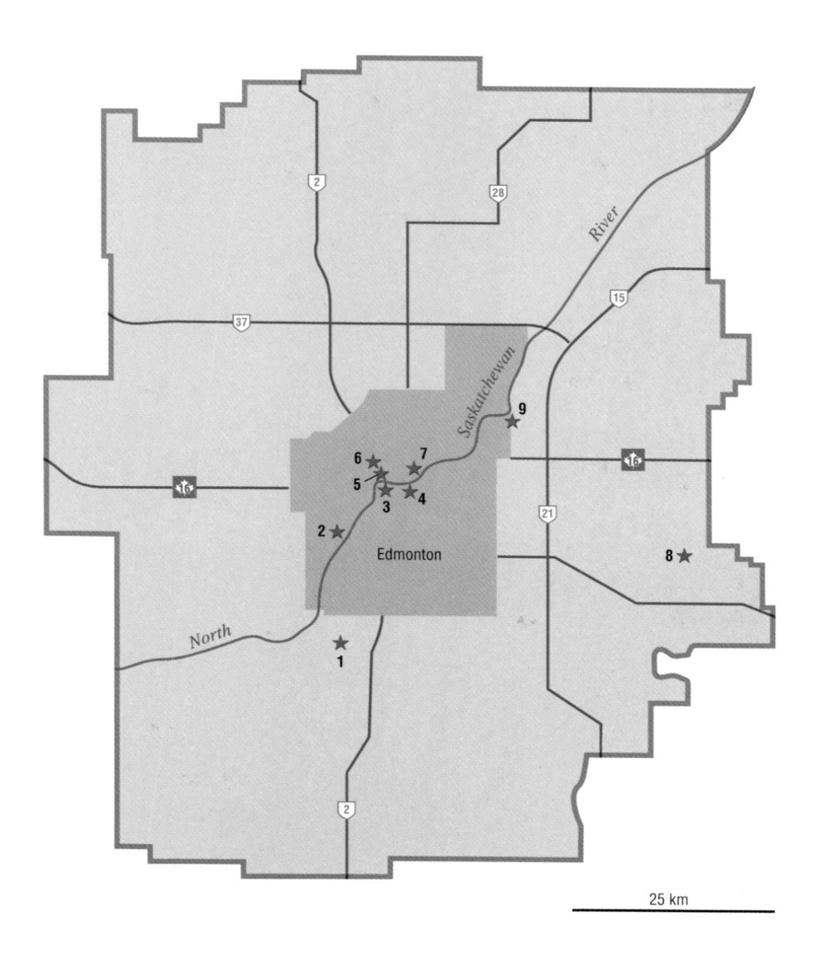


★ **1** Leduc Oil Field
★ **2** Big Bend
★ **3** Edmonton's Gold
★ **4** Mazama Ash
★ **5** Government House Park Springs
★ **6** Provincial Museum of Alberta
★ **7** Grierson Hill
★ **8** Cooking Lake Moraine
★ **9** Petrified Wood

# LEDUC NO. 1 AND THE CANADIAN PETROLEUM INTERPRETIVE CENTRE

## HIGHLIGHTS

On February 13, 1947, the discovery of oil at the Leduc No. 1 drillsite ushered in the modern oil and gas era in Canada and initiated decades of oil exploration in Alberta. The economy of the province rapidly changed from one based on agriculture to one based on petroleum, and Alberta is now acknowledged as the "energy storehouse of Canada." The Leduc field, actually a Devonian stromatoporoid reef, has produced over 480 million barrels of oil, making it one of the most productive fields in Alberta. An excellent interpretive centre has been erected one kilometre south of Devon (on Highway 60) to commemorate the historical significance of this well. The Leduc No. 1 has double recognition in that it is a Provincial Historic Site and a National Heritage Treasure.

*Ron Mussieux – Provincial Museum of Alberta*

*Highway sign welcomes visitors to the Leduc #1 Wellsite and the Canadian Petroleum Interpretive Centre.*

## THE STORY

During the Middle and Late Devonian Period, Alberta was covered by a warm, shallow ocean. Reef growth was extensive, likely because lime-secreting, sponge-like organisms, called stromatoporoids, had become extremely abundant. These sedentary creatures formed reefs that covered hundreds of square kilometres of ocean floor and often reached thicknesses of 300 metres! After millions of years, however, the shallow sea deepened and these reefs were buried under marine muds which, over time, turned into shale.

Stromatoporoid reefs are very porous and permeable and therefore make excellent oil and gas traps. The raw materials for oil and gas, technically called petroleum, are microscopic marine plants and animals. In the ancient ocean, their remains sank to the seafloor and became trapped and buried in the pore spaces between sediments. The combination of heat plus the weight of the accumulating sediments changed the organic ooze into petroleum, which migrated into the nearby porous stromatoporoid reefs. The overlying impermeable shale, called the caprock, acted as a barrier that prevented the petroleum from escaping to the surface.

The Leduc No. 1 discovery well, drilled by Imperial Oil, triggered the search for oil in other subsurface Devonian reefs in Alberta. Before long, geologists established that the Leduc reef was part of a now famous chain of reefs, called the Leduc-Woodbend "Golden Trend," many of which were also prolific petroleum producers. Thus in Alberta, Devonian reefs hold 66 per cent of our conventional oil reserves and 20 per cent of the natural gas reserves.

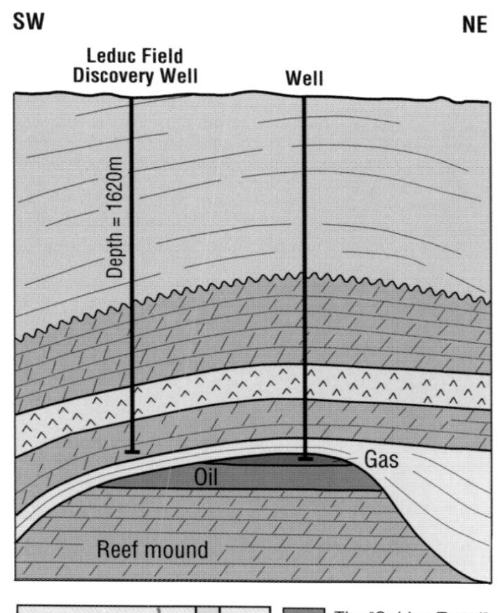


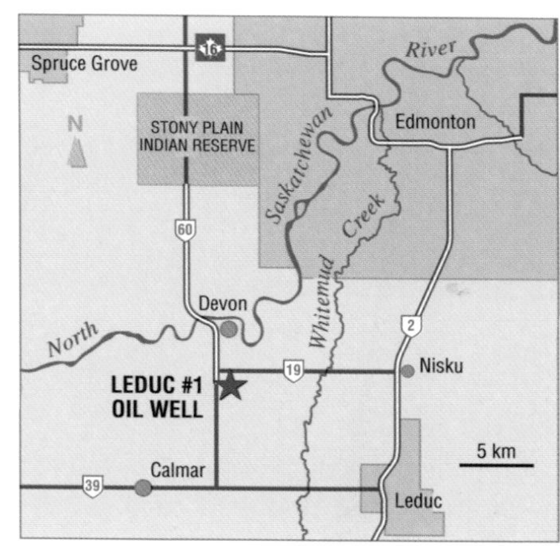


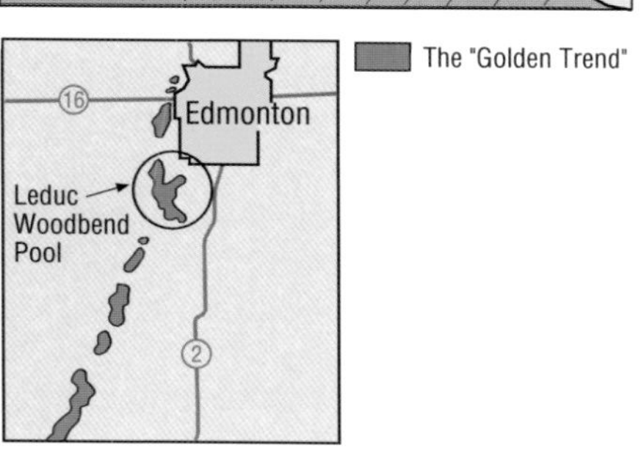

# THE CLIFFS ALONG "BIG BEND" REVEAL EDMONTON'S GEOLOGICAL HISTORY

*Ron Mussieux – Provincial Museum of Alberta*

*Over 50 metres of outcrop is exposed along the Big Bend section below Lessard Drive.*

## HIGHLIGHTS

Immediately south of Lessard Road in west Edmonton, along a loop of the North Saskatchewan River, lies a cliff of sediments that reveal much of the city's recent geological history. The sediments tell a story of a buried river channel, deposits of a 1.5-kilometre thick glacier, and the submersion of the land beneath a glacial lake. This site, locally called "Big Bend," is visited annually by geology students and can be reached by paths down to the river or can be seen panoramically from across the valley at Terwillegar Park.

## THE STORY

The cliff at Big Bend is composed of three major layers of sediments that lie on the comparatively much older Edmonton Group Cretaceous bedrock. This grey clay-rich bedrock is exposed at water level and is the same rock formation that yields dinosaur fossils here and in the Drumheller area. However, in Edmonton most of the bedrock is covered by recent river deposits and landslides.

The lowermost, and therefore oldest, layer is called the Saskatchewan sands and gravels. These were deposited over 22,000 years ago in the channel of an ancient river. The rock fragments are composed mostly of sandstone and quartzite but small amounts of petrified wood, coal, and flakes of gold are present. In this layer you can see pebbles with their long axes all oriented in the same direction — this indicates the direction the water was flowing. Similar gravel is quarried locally for the concrete industry.

Above the sand and gravel layer is a prominent yellow-brown cliff composed of fine clay and angular rock fragments. These sediments were once part of the glacier that covered Edmonton over 12,000 years ago but as the ice melted, they were dumped and left behind. The glacier was flowing towards the southwest and because many of the rocks are igneous and metamorphic, we know it came from as far away as the Canadian Shield. This type of deposit, called glacial till, has caused problems for municipal engineers because it often contains pockets of water-saturated sands which affected construction of the underground LRT tunnels.

The uppermost layer was deposited by a temporary lake known as Glacial Lake Edmonton. This short-lived lake was formed when the drainage of the North Saskatchewan River was blocked by ice near Fort Saskatchewan. In this layer, you can see thin alternating beds of silt and clay, which accumulated on the bottom of the lake. The darker clay-rich bands were deposited during the winter when the lake was ice-covered and the lighter silty beds were laid down when the lake was free of ice. Occasionally, a sheet of lake ice would carry a boulder out into the middle of the lake and drop it into the fine-grained sediments on the lake bottom. Geologists describe these "dropstones" as having been "rafted" into position.

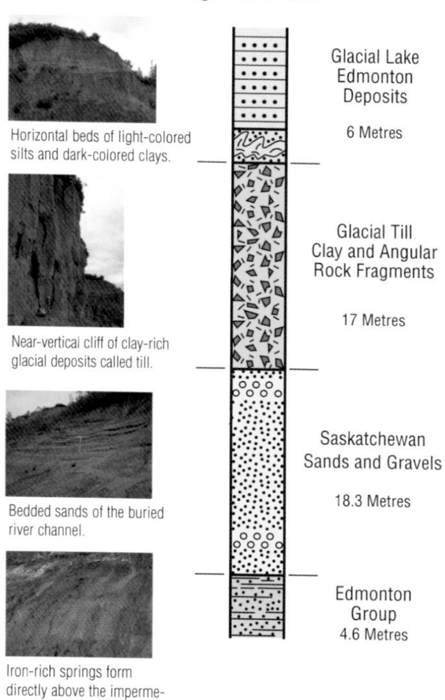


Horizontal beds of light-colored silts and dark-colored clays.

Near-vertical cliff of clay-rich glacial deposits called till.

Bedded sands of the buried river channel.

Iron-rich springs form directly above the impermeable bedrock.

*Photos by Ron Mussieux – Provincial Museum of Alberta*

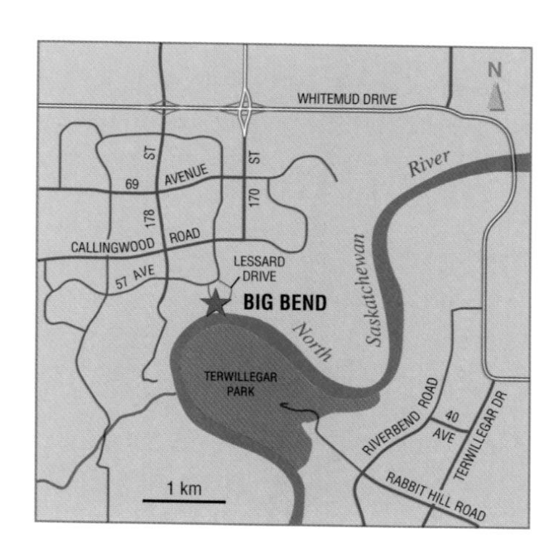

# GOLD: The Glitter in the Gravel of Edmonton's River Valley

*B.5327 – Provincial Archives of Alberta*

*At the height of gold mining in Edmonton (1895-1907), steam-powered gold dredges worked gravel bars throughout the river valley.*

## HIGHLIGHTS

In 1857, James Hector of the Palliser Expedition was the first to note gold in the gravels along the North Saskatchewan River near Fort Edmonton. This triggered a stampede of prospectors and the beginning of a local Edmonton gold rush. These gravels have been worked off-and-on since then, and today the lure of gold continues as recreational miners still pan for gold along Edmonton's river valley.

## THE STORY

Disillusioned miners travelling between the declining Fraser River and Cariboo goldfields and Fort Edmonton began working the local river valley in 1862 and, as word spread, over 300 miners arrived from across North America to try their luck. Most prospectors panned or used sluices. Rich entrepreneurs from the United States and Great Britain built steam-powered dredges to dig up the river bottom in search of this precious metal. The purpose of all these devices was to separate the gold from the other river sediments. Production reached its peak between 1895 and 1897, when 7500 troy ounces of gold were mined, or over $3 million at 1998 prices! By 1898, however, Edmonton no longer basked in the gold limelight, as miners moved north to the Klondike Gold Rush in the Yukon.

Rather than big nuggets, the gold found along the North Saskatchewan River is "flour" gold. It occurs as tiny flakes, about 0.1 to 0.5 millimetres in diameter and only a few hundredths of a millimetre thick. Unlike most gold found in nature, it is quite pure, without the usual silver content. Because the gold flakes are tiny, they are suspended in the flowing water and will only settle wherever the velocity of the river decreases dramatically, for example on the inside of a meander or a mid-channel sand bar. The sand and gravel bar east of Groat Bridge, below Emily Murphy Park, is a popular gold panning spot. Most gold is concentrated after floods and spring thaw because it is left stranded on the surface of sand and gravel bars during high water times. Thus, it is best to pan for gold during low water periods.

In addition to the modern river gravels, gold is also found locally in ancient buried river channels. Therefore, even in gravel pit operations, such as at Villeneuve, 30 kilometres from Edmonton, gold is found and recovered.

Speculation on the source, or mother lode, of the gold is ongoing. Although it is still unclear where it comes from, we do know that the richest gold areas are from south of Stony Plain to northeast of Fort Saskatchewan. Likely, the gold was eroded from local bedrock and redeposited by the river in its sand and gravel bars.

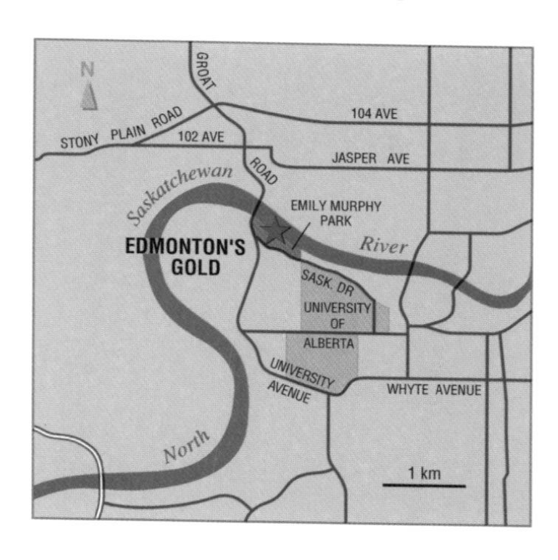

*University of Alberta*
*Flour gold of the Edmonton area. The gold is flattened and the edges are well-rounded indicating it has travelled a great distance or it has been reworked from an older deposit.*

*Dixon Edwards – Alberta Geological Survey*
*The gravels of the North Saskatchewan River have been worked for gold since the 1850s. Today, recreational gold panners still work these gravel bars.*

# MAZAMA ASH IN EDMONTON: Evidence of a Catastrophic Volcanic Explosion

*Ron Mussieux – Provincial Museum of Alberta*

*An access road near the LRT Bridge exposes silts and clays of the river's flood deposits. A thin layer of white Mazama volcanic ash seen in these sediments was deposited 6800 years ago.*

## HIGHLIGHTS

The Mazama Ash layer found along Edmonton's river valley records one of the most catastrophic events on Earth since the end of the last Ice Age. The volcanic eruption of Mount Mazama, the site of Crater Lake in Oregon some 1450 kilometres to the south, released an ash plume that blanketed almost 1.3 million square kilometres, including the southern half of Alberta. Locally, the most prominent and accessible ash outcrop can be seen just upstream from the LRT bridge on the south bank of the river. This is one of the most northerly deposits of Mazama Ash in Alberta.

## THE STORY

The building of Mount Mazama, in southern Oregon, began around 400,000 years ago. The mountain was actually not a single volcano, but rather a cluster of overlapping ones reaching as high as 3658 metres above sea level. For most of its history, Mazama's eruptions were not particularly violent. Then around 6800 years ago, it erupted explosively and sent ash high into the stratosphere with a force 100 times greater than the Mt. St. Helens eruption in 1980! The ash blew northeastwards while areas to the south and west of the volcano were relatively untouched. Throughout the entire northern hemisphere the ash would have produced spectacular sunsets.

The eruption drained the magma chamber which fed the volcano and the mountain collapsed inwards, ultimately creating a huge crater, called a caldera, that was 1219 metres deep and 9 kilometres wide. By the end of the eruption,

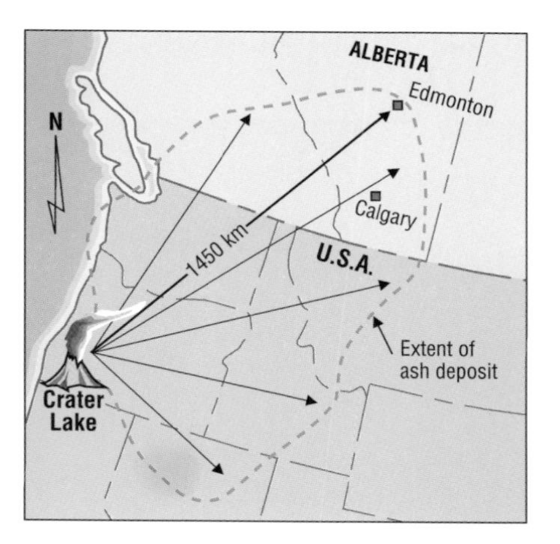


almost 70 cubic kilometres of the mountain had dispersed. Today, this caldera holds Crater Lake, the second deepest lake in North America.

At Edmonton, the ash is preserved in flood plain deposits which consist of alternating layers of dark organic-rich beds and grey silt. The ash can be found halfway down the lowermost terrace in the river valley. It is a silty pinkish-white layer and has a gritty texture because of its high volcanic glass content. The ash layer thins and thickens, likely because it was reworked by surface water after it settled. Here, it is about one centimetre thick, but in southern parts of the province it is thicker, simply because these areas were closer to the eruption. Its age was determined by radiocarbon dating of wood that was buried in the ash.

Mazama Ash is often found in exposures along river valleys in southern Alberta and like all volcanic ash deposits, it has its own distinctive mineral composition, which makes it an important tool for correlating and dating deposits across large areas. It is a particularly important tool for archaeologists studying Alberta's past because they use it to date remains of older cultures. Any artifact found below the ash is considered quite rare.

*Ron Mussieux – Provincial Museum of Alberta*

*Close-up of the Mazama Ash.*

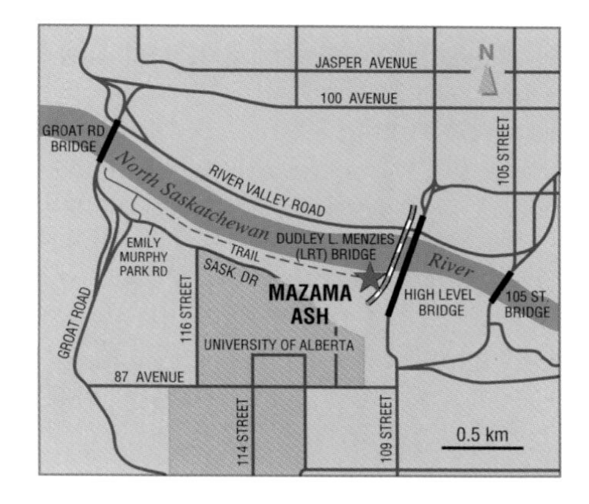

# MINERAL SPRINGS OF GOVERNMENT HOUSE PARK, EDMONTON

*Ron Mussieux – Provincial Museum of Alberta*

*These colorful spring deposits consist of hard white tufa, a calcium carbonate; soot-like black wad, a manganese oxide; and soft, red limonite, a mixture of iron oxides.*

## HIGHLIGHTS

Springs occur in a number of locations throughout Edmonton. Some of the most accessible are found along the north bank of the North Saskatchewan River valley, just at the foot of Government House Park hill. North of the bike path below the hill is a vividly colored outcrop of mineral deposits that lies partially hidden in the trees. These red, white, and black minerals are formed by springs as they trickle onto the surface from an older, buried river valley. These mineral deposits make excellent pigments, and may have been used in the past by local Aboriginal People.

## THE STORY

A spring is a place where subsurface groundwater flows from rock or soil onto land or into surface water. The Government House Park springs are formed at the boundary between a permeable gravel and sand layer and the lower clay-rich impermeable bedrock. Thus, they are called "contact springs."

The gravel and sands were deposited before the last glaciation by an ancient river that once flowed across Alberta. When the glaciers advanced over the land and occupied the Edmonton area, this broad preglacial river valley was filled in with and unsorted and unstrati-

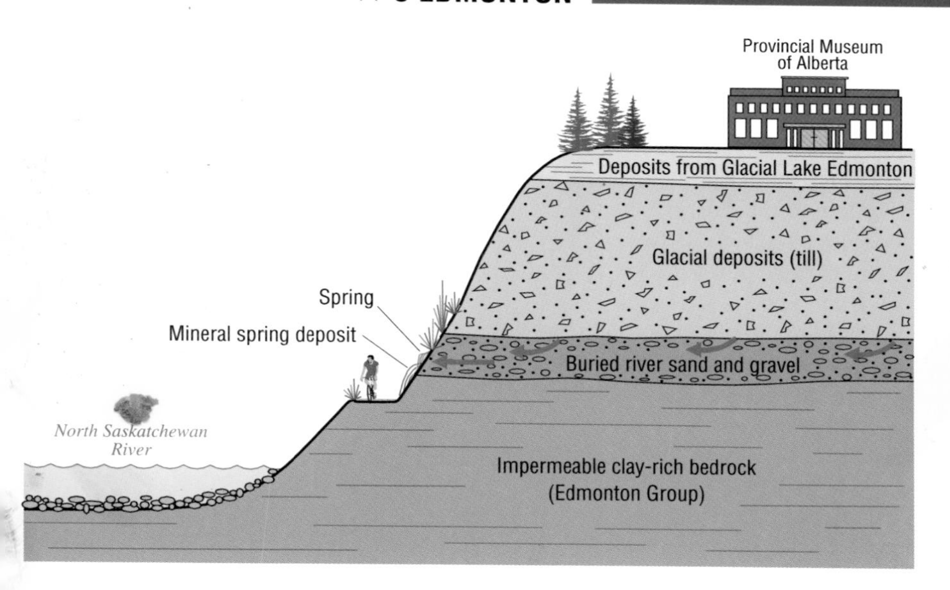


fied sediment containing rock fragments ranging in size from clay to boulders. Since the final retreat of the glaciers, the modern North Saskatchewan River has eroded its course down through these old glacial deposits to expose the sands and gravels of its ancient ancestor. Groundwater now flows along these old buried river sediments but can not pass into the underlying impermeable Edmonton Group bedrock. Where the old river channel is exposed along the banks of the modern river valley, water trickles onto the surface of the land — these are the Government House Park springs. These springs are formed in many places where the North Saskatchewan River valley intersects a buried preglacial river channel.

Springs can also be seen quite well below the Provincial Museum on a terrace of the North Saskatchewan River. They are located where the upper and lower trails join. Although the water discharge is minimal, they are marked by striking red, white, and black mineral deposits. The red deposits are formed from iron oxide minerals, while the carbonate. The black mineral, called wad, is composed of manganese oxide minerals. These minerals are forming because the spring is fed by groundwater which has seeped down through mineral-rich rocks and sediments. When the cold water comes in contact with the air, the rise in temperature and the activity of microorganisms causes the minerals in the spring water to precipitate and form a solid deposit.

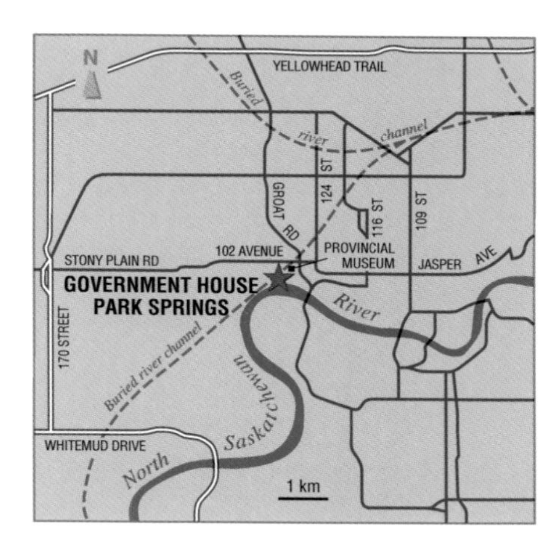

# URBAN GEOLOGY: Building Stone of the Provincial Museum and Government House

## HIGHLIGHTS

Although the Provincial Museum of Alberta houses one of the best mineral displays in western Canada, the building stones found on the museum grounds are also spectacular. Rocks from all over Canada, representing the three rock classes (igneous, metamorphic, sedimentary), can be found in the Government House and museum structures. Building stones from several provinces were used both inside and outside to reflect the confederation theme.

*Ron Mussieux – Provincial Museum of Alberta*

*Golden-brown Paskapoo sandstone of Government House is easily carved. The sidewalk is made from "Rundle Rock" siltstone quarried near Canmore.*

## THE STORY

Sedimentary rocks are the most abundant building stones at the museum. Most of the museum's outer wall is composed of Tyndall Limestone which, because of its beauty and durability, is used in many buildings across Canada, such as the Parliament Buildings in Ottawa. This limestone was deposited around 445 million years ago in a shallow inland sea and today is mined at Tyndall, Manitoba. The striking appearance of the limestone is created by the golden brown mottling against a creamy background. This mottling occurred when some parts of the limestone altered to darker dolomite. Because the dolomitized areas have tube-like shapes and branching patterns, geologists think that they represent the burrows of ancient marine worms. This stone is also rich in fossils that include corals, snails, algae, clams, and squid-like creatures, some of which are quite apparent in the polished "Tyndallstone" slabs at the museum's front entrance.

Ron Mussieux – Provincial Museum of Alberta
*A 445-million-year old snail in Tyndall limestone at the museum's front entrance.*

The outside sidewalks are made of black marine siltstone, another sedimentary rock. This rock, commercially called Rundle Rock and quarried near Canmore, is about 180 million years old. It displays ripple marks, mudcracks, and traces of numerous types of crushed fossil animals, such as ammonites. The siltstone is an easy rock to mine and use because it consists of thin beds and has regular vertical joints. Therefore the rock has a predictable, rectangular breaking pattern. Rundle Rock can also be seen in the Banff Springs and Kananaskis hotels.

Government House is built of the erosion-resistant, buff-colored sandstone of the Paskapoo Formation. This was Alberta's most important building stone and was mined mainly from quarries near Calgary. It was deposited in ancient deltas and flood plains and it outcrops as a broad belt paralleling the Rocky Mountains in west-central and southern Alberta. This sandstone is also the major building stone of the Legislature Building in Edmonton.

Other notable building stones found on the museum grounds are the glacial erratics used in the "fieldstone" walls. These erratics are a mixture of all three rock classes and were often the only building stone available for many early buildings in Alberta. The walls of the museum's main lobby are mottled white and green marble, a metamorphic rock from Ontario, and the floor is gabbro, a dark-colored igneous rock. Look for the polished granite making up the fountain and pool on the south lawn.

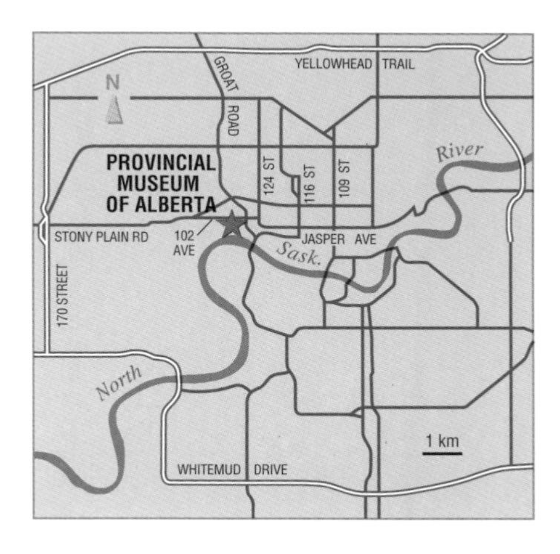

# GRIERSON HILL SLIDE, EDMONTON

*Ron Mussieux – Provincial Museum of Alberta*

*The Grierson Hill Slide is east of the Shaw Conference Centre. Today, the slide area is kept as parkland and the toe of the slide has been armored by boulders to prevent the river from reactivating the slide.*

## HIGHLIGHTS

In Edmonton, heavy rainfall on saturated ground makes surface runoff the major agent of erosion. This runoff triggers numerous earthflows and landslides along the river valley. The Grierson Hill slide is the most costly example of this type of erosion. This slide is an ideal site to look at erosional processes at work in Edmonton, how human activities have altered the speed and extent of geological change, and how we have attempted to intervene and stop further change. Grierson Hill is located on the north side of the North Saskatchewan River above the Shaw Conference Centre and just east of the Hotel Macdonald.

## THE STORY

The history of the Grierson Hill slide and the factors that contributed to such massive slope failure are well documented. City engineers use the information gained from it to plan new slope stability measures along Edmonton's river valley.

Several factors contributed to slope failure at Grierson Hill, indeed, along the entire North Saskatchewan River valley through Edmonton. First, rapid river erosion has created steep wall slopes averaging between 30° and 40°along much of the valley, and slopes up to 60° on Grierson Hill. Second, further erosion by the river is removing the foot of the slope, which adds to the instability. Third, bedrock exposed just above river level contains several thin layers of bentonite. Bentonite is a type of clay that, when it is wet, becomes slippery and turns into a perfect lubricant for local slides. Fourth, springs along the valley walls provide additional moisture which further destabilizes the slope. Lastly, human activities have probably been the final trigger to an already unstable situation, especially around Grierson Hill. These activities have included coal mining; dumping of garbage and fill on the steep slopes; the construction of buildings, roads, and parking lots too close to the crest of the valley wall; and construction of storm sewers that discharge large volumes of water into the valley.

From the early 1900s to present day, by-laws have been passed by city council to help protect and stabilize Grierson Hill from further slumping. Extra caution was taken when the Shaw Conference Centre was built in 1980. A special retaining wall was put in below the Centre and steel cables that cross under Jasper Avenue anchor the building. The bank is now armored with boulders to prevent the river from eroding the toe of the slope.

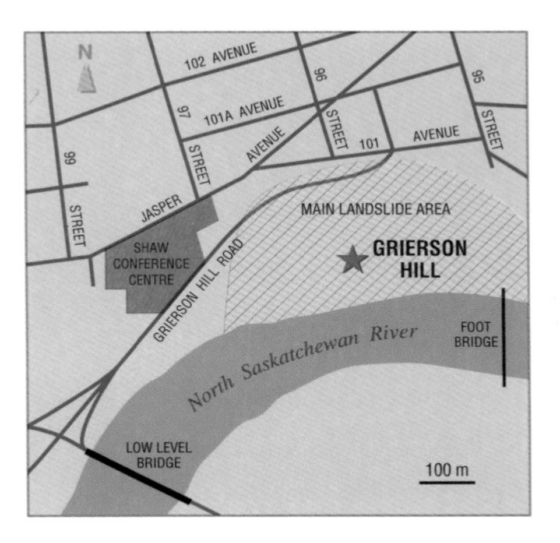

B.9282 – Provincial Archives of Alberta

*Large fractures and displaced buildings and fences resulted from significant movement of the Grierson Hill slide in 1901. A major cause of the landslide was thought to be coal mining activities at the Humberstone Mine directly below the crest of the bank.*

# THE GLACIAL ORIGIN OF THE HILLS AND LAKES OF THE COOKING LAKE AREA

*Ron Mussieux – Provincial Museum of Alberta*

*The Cooking Lake Moraine, east of Sherwood Park, consists of rounded hills separated by sloughs and lakes.*

## HIGHLIGHTS

In contrast to the flat terrain around Edmonton, the Cooking Lake area is studded with round hills and small lakes. This type of terrain, called hummocky disintegration moraine by geologists, is popularly known as "knob and kettle" topography. Like many other topographic features on the Prairies, hummocky moraine was produced during the melting of the vast continental glacier that once blanketed most of Alberta.

## THE STORY

The thick ice that covered the Cooking Lake area during the last glaciation was part of the massive continental glacier that had slowly crept southwest over Alberta. The flowing glacier scoured and eroded the bedrock, and as a result, the ice was full of rock debris. By 12,000 years ago the climate had warmed and, as the glaciers melted, huge blocks of dirty ice were stranded on the relatively high ground of the Cooking Lake district.

The glacier had an uneven layer of debris on its surface and since thick debris has an insulating effect, the ice melted at different rates. Where the ice melted quickly, the rocks slid into depressions on the surface of the ice and accumulated into steep mounds and stony knobs. Where the ice was insulated, it melted much later and formed water-filled ponds called kettles. The knob and kettle topography is especially prominent in the Blackfoot Grazing Reserve and Elk Island National Park areas.

Among the more interesting landforms of the hummocky moraine around Cooking Lake are prairie mounds or "doughnuts." These are round hills that

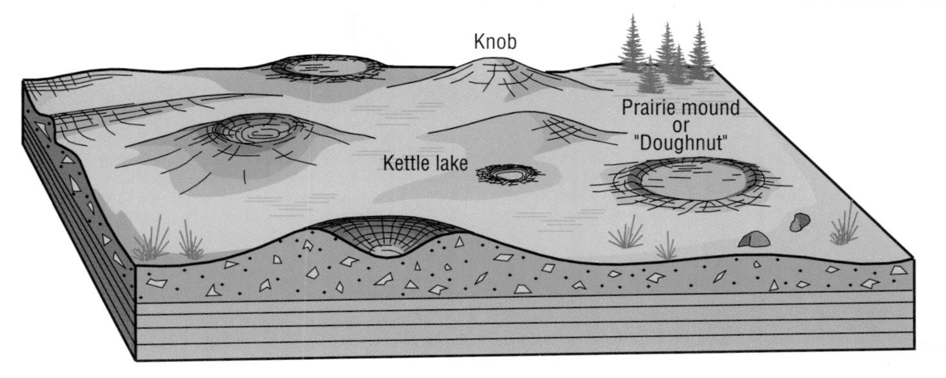


have small depressions at their peaks and actually look like doughnuts from the air. They average five metres in height and the depressions are about one metre deep. When they originally formed, they were mounds of glacial debris that had a core of ice. As the ice melted, a depression was created in the centre leaving behind the doughnut shape.

An unusual feature in the moraine directly north of Cooking Lake is a large block of bedrock that was displaced from its source some 250 kilometres away! The block is in fact a glacially transported megablock of the Cretaceous Grand Rapids sandstone, that would have been four kilometres long and twelve metres thick. Apparently during its advance southwards, the ice sheet scooped up a huge mass of the sandstone and carried it to the southwest, eventually depositing it near Cooking Lake. The power of a glacier to erode and convey rocks of all sizes is remarkable.

Hummocky disintegration moraine is a valuable environmental resource to Alberta. It is rarely used for grain farming because of the irregular landforms and poor soil, but it is used for grazing and recreational purposes. The lakes in the moraine are important resting places for migrating birds.

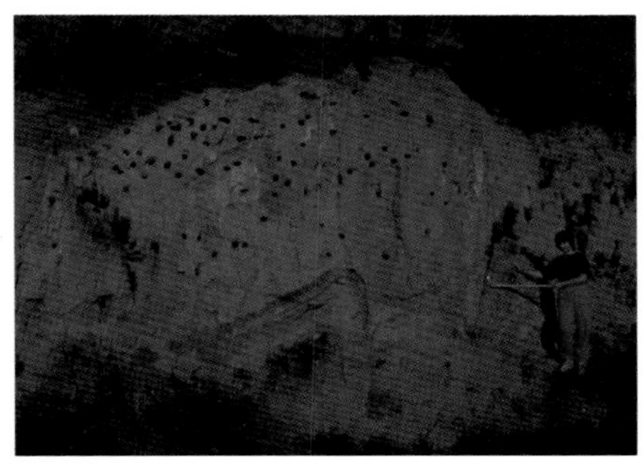

*Ron Mussieux – Provincial Museum of Alberta*

*North of Cooking Lake, the road cuts through deformed sandstone rather than the expected clay-rich glacial deposits. This sandstone is a glacial erratic and not part of the local bedrock.*

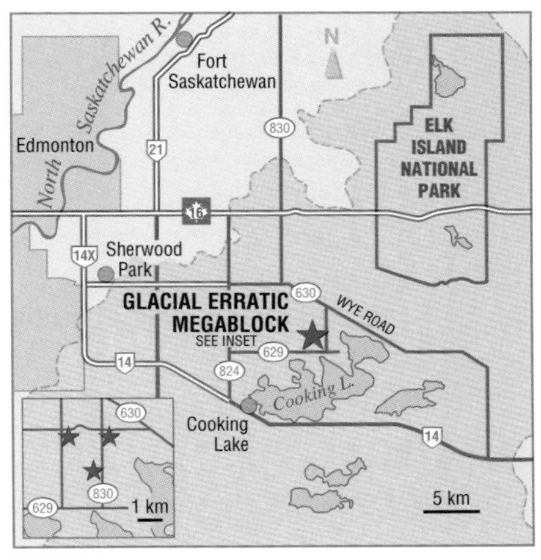


*Erratic area is shown in light brown.*

# PETRIFIED WOOD: Alberta's Provincial Stone

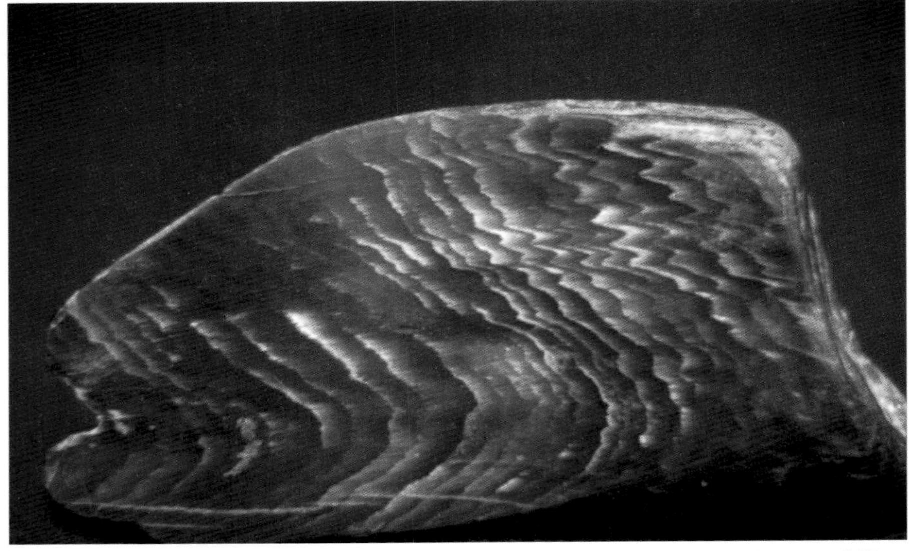

*Gregory Baker – Provincial Museum of Alberta*

*A polished slab of petrified wood from Edmonton.*

## HIGHLIGHTS

Petrified wood provides insight into the environment and vegetation that existed on our planet millions of years ago. Alberta's petrified wood is particularly popular with lapidary clubs because of its variety of structures, good range of colors, and its excellent polishing characteristics. In the Edmonton area, petrified wood can be found on gravel bars and in abandoned gravel pits along the North Saskatchewan River valley. There are several abandoned gravel pits along the river valley between Fort Saskatchewan and the community of Beverly where petrified wood is abundant.

## THE STORY

Alberta's petrified wood comes from trees that grew during the Late Cretaceous and Early Tertiary Periods, between about 90 and 60 million years ago. During much of this time, most of Alberta was covered by steamy subtropical swamps and forests. Sluggish rivers meandered through the swamps carrying large amounts of sediment that were eventually deposited on an evergrowing delta in the nearby sea to the east. This was an environment similar to the modern day Mississippi River delta. The ancient delta was covered by lush vegetation that included trees such as the dawn redwood, sycamore, bald-cypress, magnolia, and China fir. When the trees died, they were swept into the muddy rivers and soon became rapidly buried in deep mud which prevented them from rotting.

Wood is very porous. When groundwater rich in dissolved silica penetrated the buried logs, the solution was absorbed and microcrystalline quartz began to grow in the pores, replacing the plant tissue. Larger cavities in the wood were often filled with coarse quartz crystals or massive chalcedony, and the process continued until the entire log was petrified. Occasionally, sufficient individual cell detail has been retained which allows paleobotanists to identify the tree species.

The wood remained protected under hundreds of metres of sedimentary rock until the Rocky Mountains reached their greatest height in Eocene time. Since then, erosion has been slowly exposing the petrified wood along river valleys and along open expanses of badlands. The silica-rich petrified wood tumbles into nearby water courses and concentrates in rivers, stream beds, and in gravel pits throughout much of Alberta.

Gravel pits are the ideal places to find petrified wood for lapidary purposes because any softer parts have already been ground off, leaving behind only the hardest silica-rich wood. Petrified wood is made into jewellery, book ends, and numerous other products. Alberta petrified wood is commonly brown, cream, or black but occasionally reds and yellows may be found. As quartz is generally white or grey, the different colors found in petrified wood are due to impurities such as iron oxides, manganese oxides, or organic carbon.

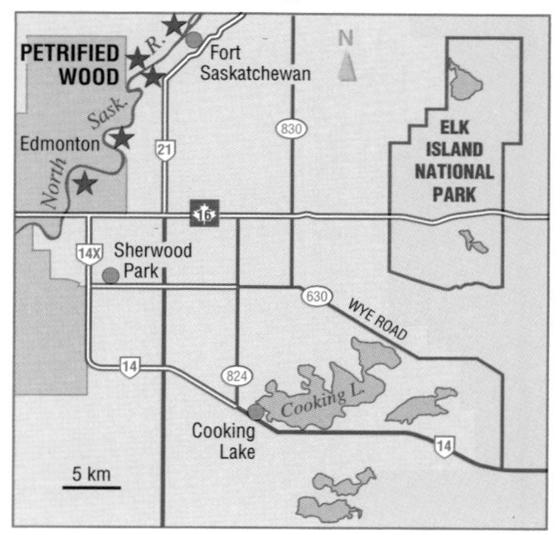

Ron Mussieux – Provincial Museum of Alberta

*Petrified wood is plentiful in the gravel pits east of Edmonton.*

# CHAPTER 4: ALBERTA'S HEARTLAND

*White marl is deposited in the muskeg near Marlboro, the site of Alberta's first cement plant.*
*Ron Mussieux – Provincial Museum of Alberta*

# ALBERTA'S HEARTLAND

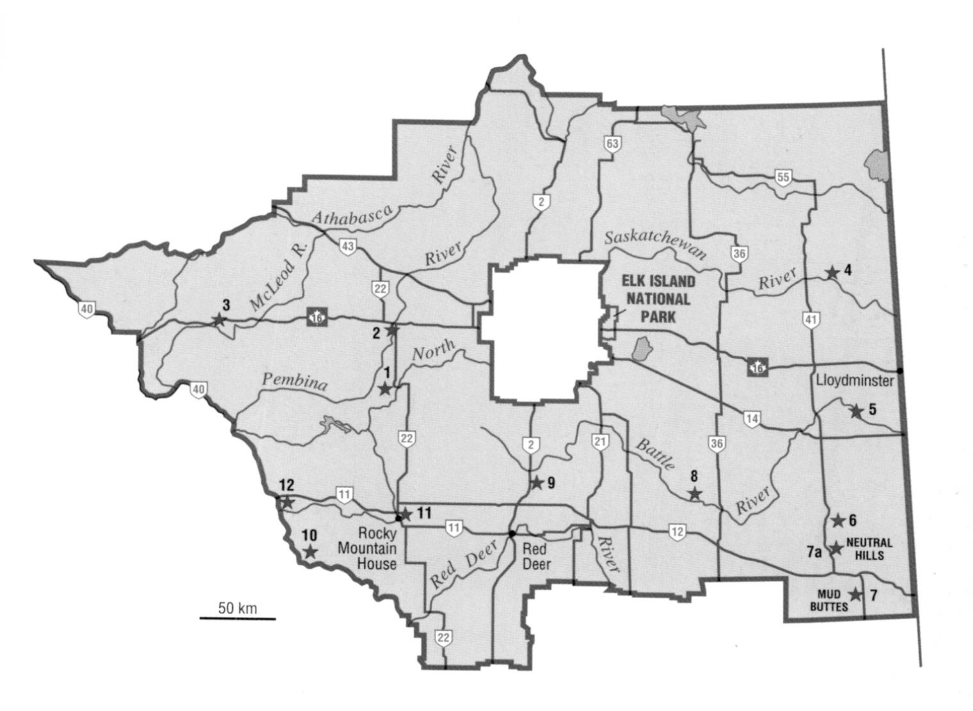


★ **1** Pembina Field Discovery Well
★ **2** Pembina River Valley
★ **3** Marlboro
★ **4** Lindbergh Salt Plant
★ **5** Koroluk Landslide
★ **6** Metiskow Plant
★ **7** Mud Buttes
★ **7a** Neutral Hills
★ **8** Diplomat Mine
★ **9** Wolf Creek Dunes
★ **10** Ram Falls

**David Thompson Highway Roadlog**

★ **11** Road Stop 1
- Rocky Mountain House

★ **12** Road Stop 2
- Nordegg

## PEMBINA: Canada's Supergiant Oil Field

*Julie Hrapko – Provincial Museum of Alberta*

*A pumpjack, sometimes called a horsehead pump, brings oil to the surface in the Pembina oil field.*

### HIGHLIGHTS

This discovery well led to the opening of the Pembina Field, Canada's most productive oil field, which has yielded more than 228 billion litres (or 1.4 billion barrels) of oil to date. It is one of the largest oil fields in the world, extending over nearly 3500 square kilometres. Approximately 7800 wells have been drilled since its discovery in 1953, with a remaining 7000 still pumping. With Alberta supplying 85 per cent of Canada's conventional crude oil, this single oil field has made a significant contribution to this country's petroleum economy.

### THE STORY

Although we speak of oil fields or pools, oil does not lie underground in vast caverns. Instead, it is stored in tiny pores in sedimentary rock such as sandstone and limestone, similar to water in a sponge. Rock formations that hold high concentrations of petroleum are called reservoir rocks. In order for petroleum to accumulate in a reservoir rock, there also needs to be an impermeable seal to prevent it from escaping, a trap which holds it in one area, and a nearby source rock, such as shale, that was once rich in hydrocarbons.

In the Pembina Field, the major reservoir rock is the 100-million-year old Cretaceous Cardium Formation sandstone which lies below much of the Foothills and the western edge of the Plains of Alberta. This layer of porous sandstone, which is overlain and underlain by impermeable shale, slopes up towards the east and "pinches out," similar to a wedge. The oil seeps out of the surrounding shale, migrates laterally up the tilted sandstone beds, and accumulates as it reaches the overlying shale seal. This highly efficient oil trap is a classic example of a stratigraphic trap.

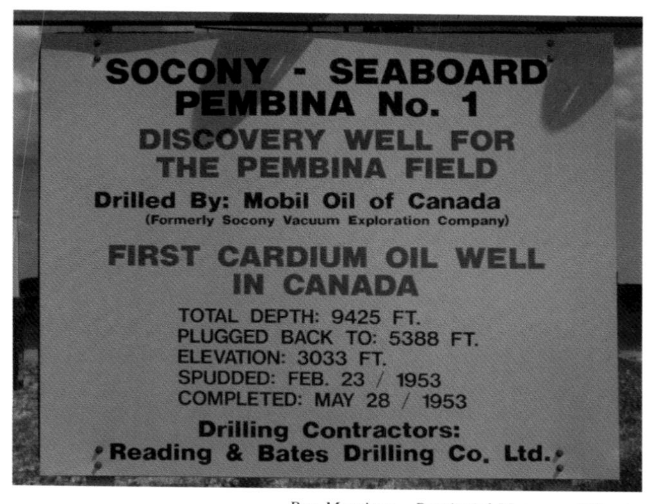


*Ron Mussieux – Provincial Museum of Alberta*

*A plaque containing historical information is attached to the Pembina #1 Discovery Well.*

Since the development of Pembina Field, the Cardium Formation has gained worldwide recognition for its enormous hydrocarbon storage capacity. The sand was deposited during the Upper Cretaceous in a seaway that extended from the modern Gulf of Mexico to the Arctic Ocean. The eroding mountains to the west provided the sand which accumulated as a curved strip nearly 1000 kilometres long. The initial volume of oil held in the Cardium Formation has been estimated at 1600 billion cubic metres, unfortunately only 20 per cent can be recovered with today's technology. This is because the sandstone is not particularly permeable, meaning the petroleum moves slowly from one pore space to another, and hence slowly into and up the drill hole. A typical Pembina well may only pump three cubic metres per day. This is different from the porous and permeable Leduc reefs where the oil flowed quickly to the surface. Nevertheless, Pembina has provided at least 10 per cent of Alberta's cumulative oil, certainly an admirable contribution to our economy.

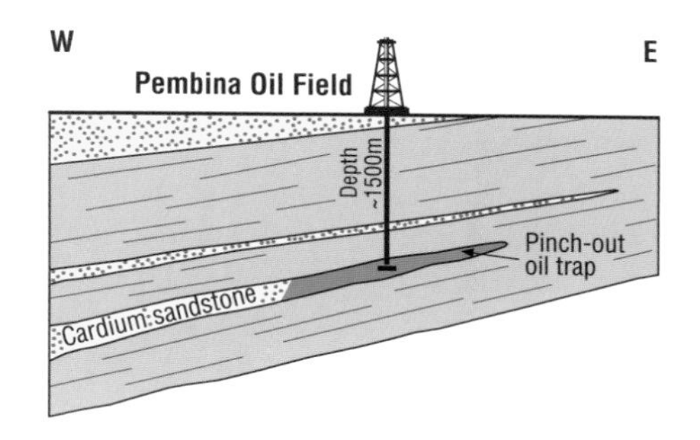


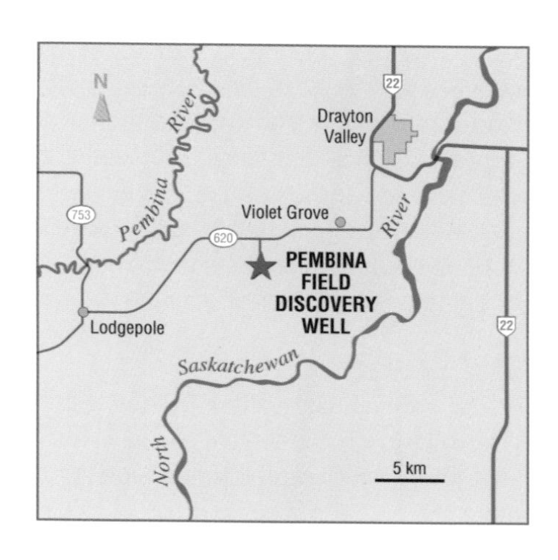

# PEMBINA RIVER VALLEY NEAR PEMBINA RIVER PROVINCIAL PARK

*Ron Mussieux – Provincial Museum of Alberta*

*Buff-colored sandstone cliffs form the upper part of the valley walls, while soft grey sandstones and shales form the lower part.*

## HIGHLIGHTS

Between Evansburg and Entwistle, the Pembina River cuts a narrow, picturesque valley near Pembina River Provincial Park. Located 100 kilometres west of Edmonton on Highway 16, this valley was once the scene of a thriving coal industry and a short-lived building stone quarry. The valley walls display a very obvious boundary, or contact, between two rock units: an upper buff-colored sandstone and softer, underlying grey sandstones and shales. Besides the color change at the contact, the angle of the slope changes from near-vertical cliffs in the buff sandstone to around 45° in the grey sandstones and shales.

## THE STORY

The spectacular Pembina River valley is geologically quite young and only began to form around 13,000 years ago, upon the retreat of the continental glaciers. The valley here is 60 metres deep and over 350 metres wide.

The lower grey deposits, called the Edmonton Group, were deposited around 70 million years ago in a broad, shallow delta at the edge of the inland sea. The Edmonton Group consists of clays and shales with some sandstone beds. The distinctive light grey color comes from the abundant bentonite (see page 222) in the beds. There are coal seams of varying thicknesses throughout the formation with the thickest ones occurring near the top. The uppermost "Big Seam" or "Pembina Seam" has a thickness of six metres of low rank sub-bituminous coal. It occurs beneath the river here but outcrops farther north in the valley and is being actively mined at Wabamun Lake in huge open pit mines.

Above the Edmonton Group lies the massive yellow sandstones of the Paskapoo Formation. This formation is younger than the Edmonton Group and was deposited in deltas and flood plains. The sandstone is very resistant and in this area forms cliffs around 23 metres high. Sedimentary structures, like cross-bedding and ripple marks, together with plant fragments are rare but can be found here.

This area was important at the turn of the century because of the extensive coal deposits. Several mines operated, such as the underground Pembina mine at Evansburg and some open pit mines farther north. Besides the coal, the massive cliff-forming sandstone of the Paskapoo Formation was considered of importance as a building stone. In the early 1900s, Pembina Quarries Limited, of Edmonton, began mining the Entwistle area adjacent to the railway tracks. From here, the stone was shipped to construction projects in Edmonton. However, the stone turned out to be of poorer quality than the Paskapoo sandstone from Calgary and the Entwistle operations were soon shut down.

The Provincial Park campground located here has numerous walking trails which provide excellent views of the valley.

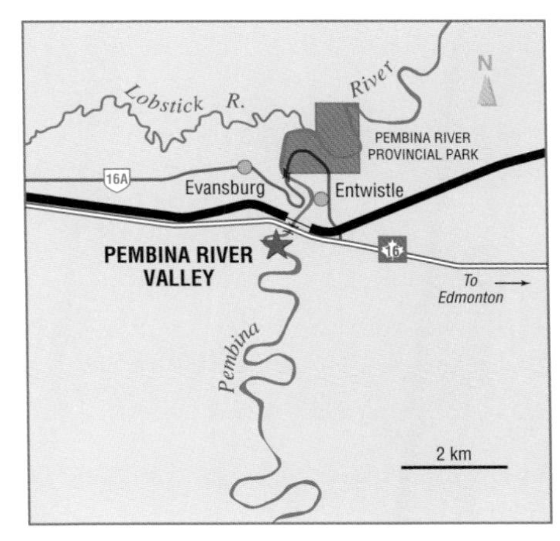

*Ron Mussieux – Provincial Museum of Alberta*
*Abandoned sandstone blocks in overgrown quarry.*

*B.1373 – Provincial Archives of Alberta*
*The sandstone quarry in operation at Entwistle.*

## MARLBORO: Lonely Chimney in the Muskeg

*Ron Mussieux – Provincial Museum of Alberta*

*The chimney stands above the ruins of the Marlboro Cement Plant.*

### HIGHLIGHTS

Many motorists travelling the Yellowhead Highway (Highway 16) have pondered why a lonely chimney rises 56 metres above the muskeg near Marlboro, 35 kilometres west of Edson. This is the site of Alberta's first cement plant and the smokestack is a reminder of a dream that fell far short of its goals. The plant was built in 1913 by the Edmonton Portland Cement Company. This site was chosen because of the nearby extensive marl and clay deposits and its closeness to the Grand Trunk Pacific Railway. Marl was once used as a major constituent in cement.

### THE STORY

Marl is a soft, crumbly, fine-grained limestone (calcium carbonate) that forms only in fresh waters. The Marlboro region has areas of high relief, and as groundwater moves through these higher hills it dissolves limestone fragments in the glacial deposits. This lime-rich water trickles out as springs near the sloughs and ponds adjacent to the plant. As the spring water warms, it can not hold as much carbon dioxide and calcium carbonate (marl) is precipitated. The precipitation of marl is further increased by animal and plant activities that change the carbon dioxide and calcium concentration in the water.

Marl can be used as a livestock feed supplement, as a treatment for acidic soils, or in cement. Cement is composed of calcium carbonate and silica. At very high temperatures, these two fuse together to form solid blocks. The blocks are then crushed into a fine powder that we call cement. This is mixed with water, sand, and gravel to form concrete.

At the beginning of this century, Edmonton was going through a major development boom. Projects like the High Level Bridge and the Legislature Building required large amounts of cement, which had to be shipped in from Ontario at high cost. When marl was discovered by railway workers at the Marlboro site in 1910, the people of Edmonton thought they had found an inexpensive alternative to the Ontario cement. With a large deposit of marl and nearby silica and coal to heat the kilns, the plant was expected to be a great success. Well known and wealthy Edmonton businessmen, like George Bulyea and Henry Marshall Tory, invested a total of three quarters of a million dollars in the cement plant. Unfortunately, the plant never did as well as expected because the nearby silica was unsuitable and there was too much competition from other cement plants. The plant was closed in 1931 and today all that remains are the towering smokestack and the foundation walls of the old buildings that are now being overgrown by trees. The marl, which gives the nearby town of Marlboro its name, still forms and can be seen clearly from the edge of the highway.

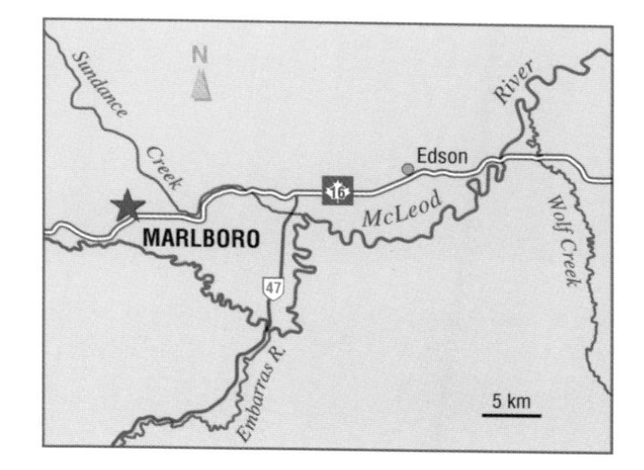

*B.1370 – Provincial Archives of Alberta*

*The Edmonton Portland Cement Company Plant at Marlboro in the 1920s.*

# LINDBERGH SALT PLANT: From Salt Water to Rock Salt

*Ron Mussieux – Provincial Museum of Alberta*

*The Lindbergh Salt Plant, north of Vermilion, produces 400 tonnes of salt per day.*

## HIGHLIGHTS

Most of the Windsor-brand table salt you use to season your food or melt ice on your driveway is actually not mined in Windsor, Ontario, but in Lindbergh, Alberta. In 1946, companies drilling for oil and gas in this area accidently discovered thick beds of rock salt. The Lindbergh salt plant, located 65 kilometres north of Vermilion, is one of two operating salt mines in Alberta and currently produces about 400 tonnes of halite, or rock salt, per day.

## THE STORY

During the Devonian Period, some 375 million years ago, rising sea levels sent sea water flooding across much of Western Canada. This shallow arm of the ocean, called the Elk Point Basin, was bounded on the east by the Canadian Shield and on the west by the Western Alberta Arch, a long, narrow peninsula that extended from Montana to the Peace River District. These warm, tropical waters teemed with marine life and both corals and stromatoporoids established reefs in many parts of the basin.

A rock arch near the south shore of what is now Great Slave Lake left the water shallow enough for stromatoporoids and corals to build the Presqu'ile Barrier Reef. The reef blocked circulation of sea water from the open ocean into the shallow basin and in the arid climate, water evaporated and the salinity increased until most sea life was killed. Special minerals, called evaporites, came out of solution and settled as a layer on the basin floor. A 100-metre column of sea water will produce only 0.5 metres of evaporites and in Alberta, evaporites, mainly rock

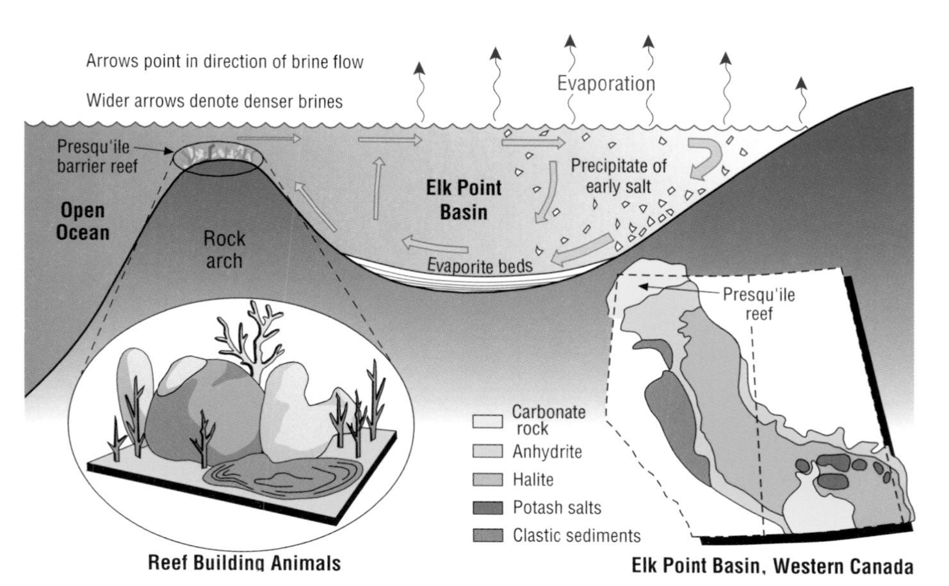


Reef Building Animals

Elk Point Basin, Western Canada

salt, reach thicknesses of over 400 metres. This means that in order to reach such thicknesses, fresh supplies of sea water must have flooded into the basin numerous times. Evaporites form in an ordered sequence — first to form are the sulphates (gypsum and anhydrite), then halite, and finally the bitter dregs of potassium salts (potash). The Saskatchewan portion of this basin contains over 50 per cent of the world reserves of potash which is used in the manufacture of fertilizer.

At Lindbergh, the salt is located over 1100 metres below the plant and is brought to the surface by a technique called "solution mining." This is done by drawing nine million litres of water per day from the North Saskatchewan River and then pumping it down wells to the salt beds. The water warms as it descends, dissolves the salt and is then pumped up another well. Each 10 litres of water carries over 2.75 kilograms of salt, which is pumped into large pans and steam-heated in evaporators until 99.8 per cent pure salt is left.

The Lindbergh plant produces fine table salt, food-processing salts for meat packing, bakeries and fisheries, chemical ice control (safety) salt, water conditioner salt, and agricultural block salt. The company believes there is enough salt beneath Lindbergh to last us another 2000 years!

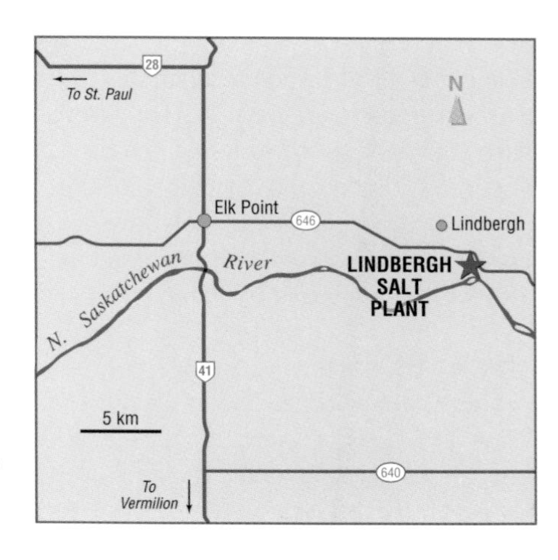

# THE KOROLUK SLIDE: A Slope in Motion

David Cruden – University of Alberta

*The steep, 14-metre scarp at the top of the Koroluk Slide was photographed in 1974 shortly after the slide occurred.*

## HIGHLIGHTS

Even on the rolling hills of the Prairies, landslides are a significant, and often destructive geological force because of the weakly cemented sediments and steepwalled valleys. Landslides are particularly active along the entire length of the Battle River, as experienced by George Koroluk who watched his land sink 14 metres in only a few weeks. Although this site is on private land, visitors are welcome and a viewpoint, interpretive plaque, and guest book have been provided. The scarp is still fresh and visibility is not impaired by trees.

## THE STORY

The Koroluk Slide began abruptly in May, 1974, as a large crack that ran across a cultivated field near the top of the Battle River valley. Within days the field began to sink and by August 28, a near vertical scarp, 14 metres high and 180 metres long, had formed. This type of slide, called a mudslide, occurs when the layer of earth just beneath the surface, the subsoil layer, becomes oversaturated with water. The heavy topsoil squeezes down on this subsurface mud which then flows downslope, still underneath the surface, as a moving lobe or tongue. Here, the mud slid 230 metres before stopping, and cracks can be seen where it raised the topsoil. As the subsurface layer disappeared downhill, the ground surface sank to replace it, creating a depression, called a graben. At the time of the slide, Mr. Koroluk's crops were planted, and were virtually undisturbed by their rapid drop.

Mudslides are often caused by excessive rainfall, especially when combined

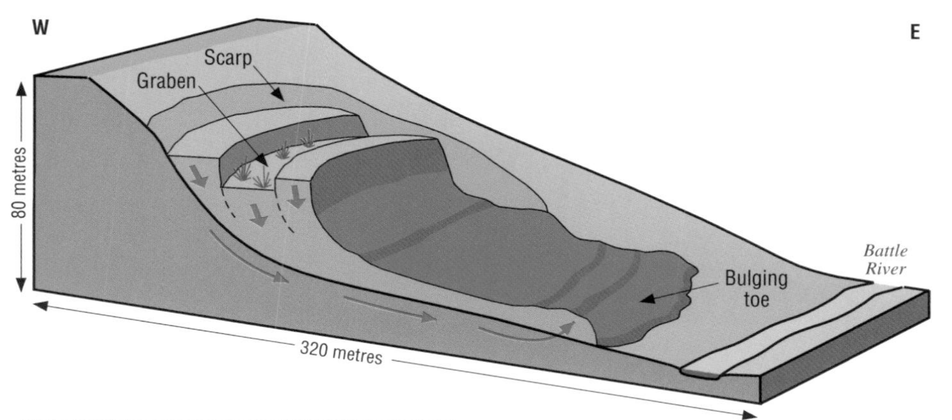

Ron Mussieux – Provincial Museum of Alberta

*View northeast from the top of the slide towards the Battle River. Notice the down-faulted portion of the slide (called a graben) that is carrying the still-living trees (photo taken 1996).*

with changes to the natural vegetation. The history of this area indicates that by 1963, a forested area behind the crest of the slope was cleared for cultivation, and this was followed by years of higher than average precipitation. The removal of the trees likely allowed for increased infiltration of rainwater and for more snow accumulation beneath the crest of the slope. As well, the soils along the Battle River are rich in clay and bentonite which, when combined with water, turn into a slippery, fluid mud. This type of earthflow is common in Alberta and must be considered when planning new roads, railways, and agricultural land. The Koroluk slide is now converging with similar slides to the south, producing a combined total of over a million cubic metres of displaced mud!

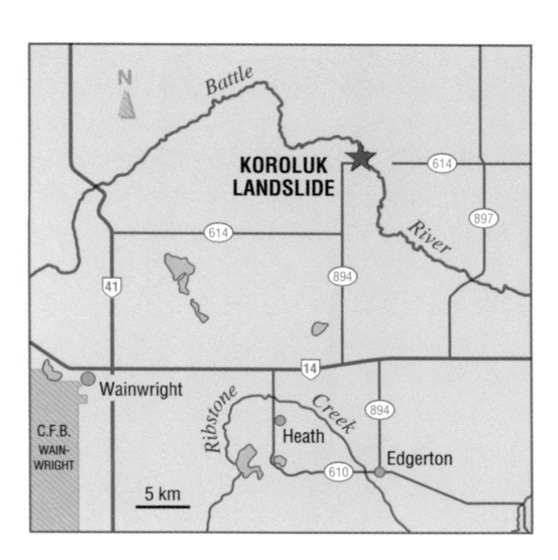

# METISKOW SODIUM SULPHATE PLANT:
## Mining an Alkaline Lake

*Ron Mussieux – Provincial Museum of Alberta*

*The Metiskow Plant on the shore of the alkaline Horseshoe Lake. This plant mined these brines for over twenty years and may reopen soon.*

## HIGHLIGHTS

Eastern Alberta between the North and South Saskatchewan river valleys consists of rolling hills dotted with sloughs and shallow lakes. Few of these water bodies drain and over time, their salt content builds up and they become brackish and bitter-tasting. Their shorelines are frequently marked by a glistening coat of white salts; these have been mined in Saskatchewan and Alberta for over 75 years and, in 1991, provided a 50-million dollar industry. Horseshoe Lake is the only Albertan deposit now known to be of commercial value.

## THE STORY

The white salt forming in these lakes and sloughs is not sodium chloride (common table salt) but primarily hydrated sodium sulphate (called Glauber's salt or mirabilite). Unlike the ancient marine chloride salt deposits being mined at Lindbergh and Fort Saskatchewan, these slough salt deposits are formed on land in these shallow water bodies. A combination of high summer evaporation, poor drainage, low precipitation, and long cold winters has produced these economically important sodium sulphate deposits.

Rain falling on nearby land surfaces leaches mineral salts from the soil, glacial deposits, and shallow bedrock. The groundwater carries these dissolved minerals to the shallow lakes and sloughs. Summer evaporation concentrates salts in these water bodies and the chilling of water during the winter results in the deposition of hydrated sulphate crystals on the lake bottom. Hydrated sodium sulphate is the mineral mirabilite, which forms large, clear, bladed crystals that can grow up to 1.5 metres in 48 hours! Unfortunately for the collector, mirabilite contains water in its crystal structure, and as soon as it is exposed to air it dehydrates and crumbles into white powder identical to that coating the lake shore.

The Metiskow Plant, on Horseshoe Lake, was in production from 1969 to 1991 and is now being upgraded and may begin production again in the future. This lake consists of four basins that are separated from each other by man-made dykes. The sodium sulphate occurs in solution in the lake water and as crystals intermixed with clays on the lake bottom. These sulphate reserves are over 17 metres thick in the south basin and the whole lake has reserves estimated at over three million tonnes. This lake has existed for over 10,250 years based on a radiocarbon date from the base of the salt deposits.

Over the years, a variety of mining techniques have been used to mine these salt beds. The present method pumps the excess brine off to another basin and then mines the dry beds during the winter with a backhoe. The mirabilite is then passed through large evaporators at the plant to drive off the water. This produces a fine pure powder of sodium sulphate, called salt cake. Salt cake is used in the production of kraft-paper, as a substitute for phosphates in detergents, and to produce fertilizer.

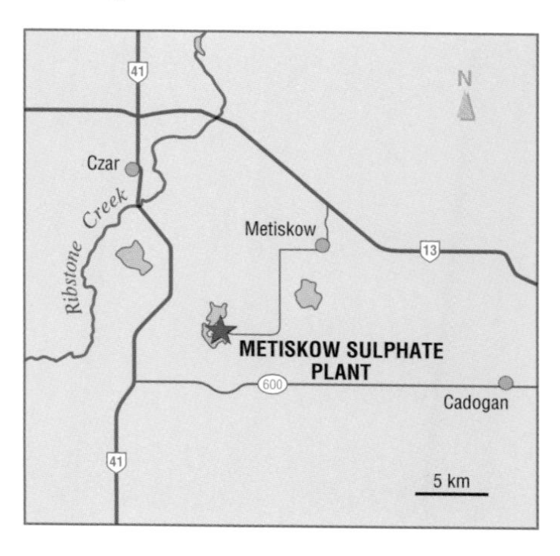


*Dog-tooth shaped mirabilite crystals grow downwards from a floating crystal raft into the alkaline-rich water.*

*William Last – University of Manitoba*

*Ron Mussieux – Provincial Museum of Alberta*

*In order to dry and mine the lake bottom deposits, a dredge on Horseshoe Lake pumps brine from one lake basin across a man-made dyke and into another basin. This lake differs from most alkaline lakes because it also has a high carbonate content.*

## MUD BUTTES AND NEUTRAL HILLS: Deformation by Glacial Ice

Ron Mussieux – Provincial Museum of Alberta

*Highly deformed sandstones and shales, Mud Buttes.*

### HIGHLIGHTS

Mud Buttes is a group of low hills 15 kilometres south of the hamlet of Monitor, east of Coronation on Highway 12. It is an isolated pocket of badlands about two kilometres long and 800 metres wide, and is probably North America's largest and best exposed site of glacially deformed bedrock. The spectacular folds and faults seen here are formed by the push from advancing glaciers, and provide excellent information about the direction of flow of glaciers during the last Ice Age. There is a picnic shelter and parking area on Mud Buttes, east of the road, but the non-gravelled trail to the shelter should be avoided in wet weather. The Neutral Hills, located north of Consort, are about 15 kilometres long, 3 kilometres wide, and 120 metres high. Although they are much larger than Mud Buttes, they were also formed by the thrusting action of an advancing continental glacier.

### THE STORY

The rocks forming Mud Buttes are weakly cemented sandstones and mudstones of the Belly River Formation, deposited around 75 million years ago during the Late Cretaceous Period. These soft rocks were bulldozed into faults and rounded, toothpaste-like folds by the immense weight of an overriding glacier. The advancing glacier tore loose large sheets of bedrock and shoved them about five kilometres to the southwest, stacking them against each other to produce Mud Buttes. Measurements of the folds and faults, plus the presence of igneous and metamorphic rocks in nearby glacial sediments, indicate that the glacier advanced into this area from the northeast from as far away as the Canadian Shield, at least

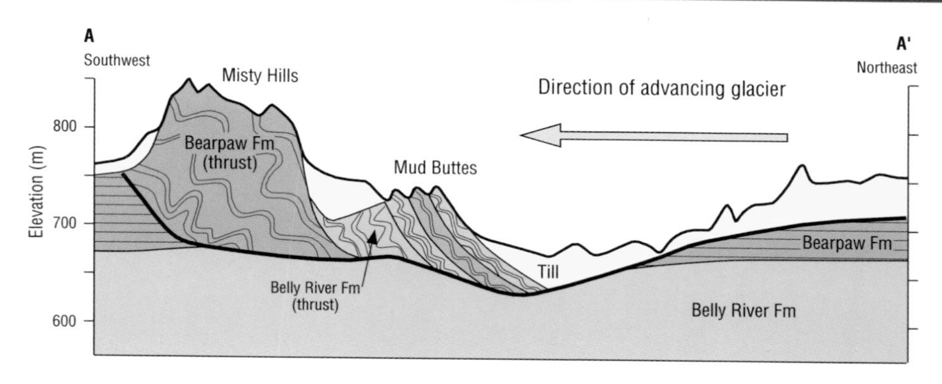


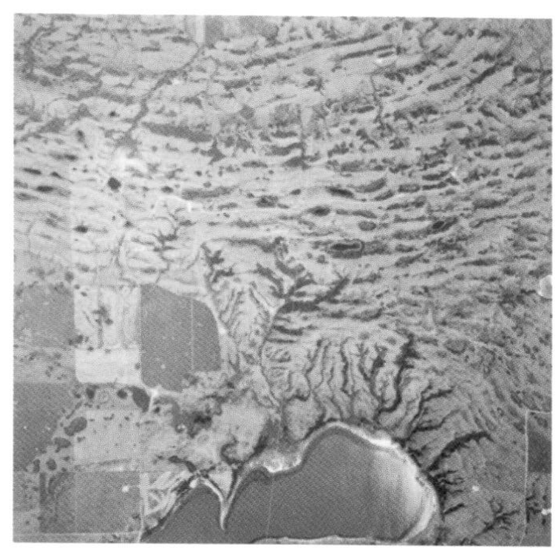

*Government of Alberta airphoto*

*An aerial view of the Neutral Hills and Gooseberry Lake Provincial Park. These hills consist of parallel concentric ridges that curve northwards, indicating they were formed by a massive south-flowing glacier.*

700 kilometres!

Rocks equivalent in age to the Belly River Formation are found in many areas of the province as flat-lying beds; for example, the fossil-rich badlands at Dinosaur Provincial Park. At Mud Buttes, however, the glacier thrust the beds up to an angle of nearly 30°. The nearby Misty Hills, Neutral Hills, and Nose Hill all owe their origin to the push from an advancing glacier.

The Neutral Hills, best viewed along Highway 41 between Czar and Consort, and the related Nose Hill, have also been formed by glacial push. The Neutral Hills have been described by geologists as a "hill-hole pair." The massive continental glacier flowing from the north scooped out a large block of soft bedrock creating a broad depression. Farther south, along the direction of ice movement, it deposited this bedrock as a series of concentric, lobe-shaped ridges of considerable height. Highway 41 extends over a low pass in the Neutral Hills, and from here the parallel nature of the ridges that form the Hills can be seen clearly.

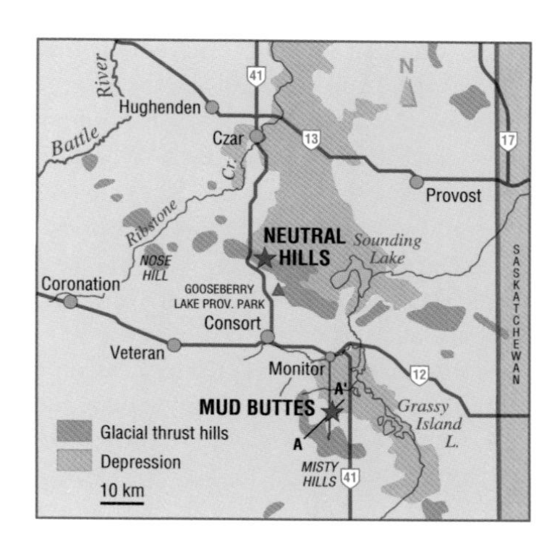

# DIPLOMAT SURFACE MINE MUSEUM INTERPRETIVE CENTRE: Open Pit Coal Mining on the Prairies

*Ron Mussieux – Provincial Museum of Alberta*

*The massive recently restored Marion 360 coal stripping shovel.*

## HIGHLIGHTS

The Diplomat Mine Interpretive Centre is a picnic site adjacent to Big Knife Provincial Park. It is an ideal place to learn about open pit coal mining and to examine mammoth coal mining equipment which normally can only be viewed from a distance. The Centre exhibits the huge Marion 360 stripping shovel which weighs about 500 tonnes. When it was built in the 1920s, it was one of the largest mobile land machines ever assembled. This shovel was restored by the Diplomat Mine Museum Society and has been declared a Historic Site by the Alberta Government. Also on exhibit are a smaller coal shovel and the 30 cubic yard bucket from "Mr. Diplomat," the mine's largest shovel. There is a raised platform that allows visitors to view an area of older, abandoned coal mine tailings. Nearby are the shutdown Diplomat Mine, the still operating Paintearth and Vesta coal mines, and Alberta Power's Battle River Generating Station.

## THE STORY

Underground coal mining began in the Battle River valley in 1906, but the Diplomat Mine, the region's first surface mine, did not start production until 1950. Surface mining, or open pit mining, is a completely different process from underground mining (see Bellevue and Atlas mine sites, pages 202 and 220).

The requirements for a successful open pit coal mine, are the presence of one or more flat-lying coal seams of economic thickness, and a thin layer of

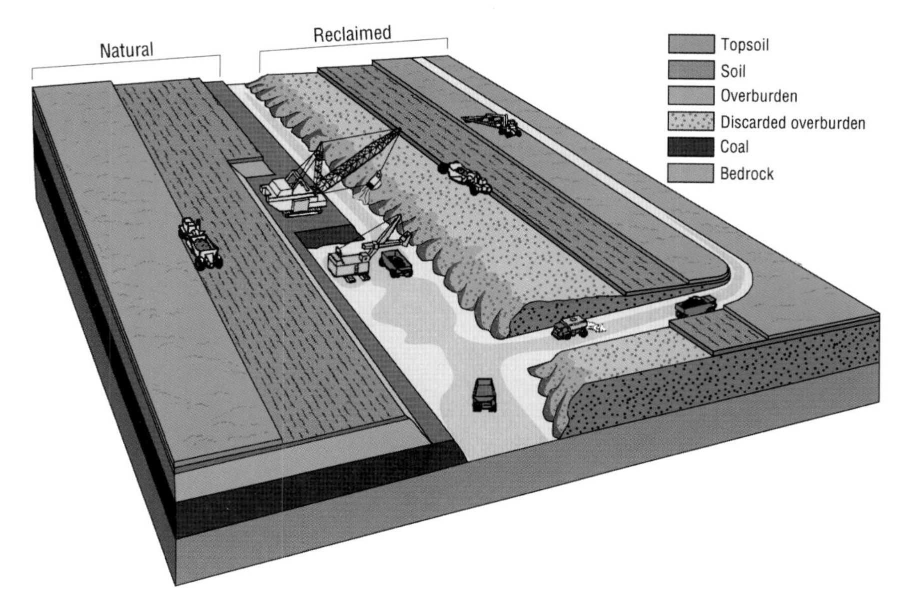


waste rock overlying the coal seams. Called overburden, this waste is stripped away by massive earth-moving equipment. The Diplomat Mine produced from a seam that was between 2 and 5 metres thick and had an overburden thickness of 5 to 12 metres. Today, both the Paintearth and Vesta mines extract coal from a similar geological setting.

The prime purpose of these three mines has been to supply low-sulphur coal to the Battle River Electrical Generating Station, a coal-fired power plant. Today, 90 per cent of Alberta's electrical power is produced in this type of plant, and this requires 60 per cent of Alberta's total coal production. In 1995 alone, Alberta produced nearly 42 million tonnes of coal — over five times as much as was produced in the coal boom of 1950! Thus, contrary to popular belief, coal mining is still a major energy industry in Alberta.

The Battle River coalfield is located on valuable agricultural land and therefore reclamation is an integral part of the mining process. Prior to mining, the topsoil is removed and stockpiled. The overburden is then stripped using large draglines and the exposed coal is removed with coal shovels and front-end loaders. Once mining is complete, the overburden material is levelled, and the topsoil is replaced, cultivated, and seeded to produce crops.

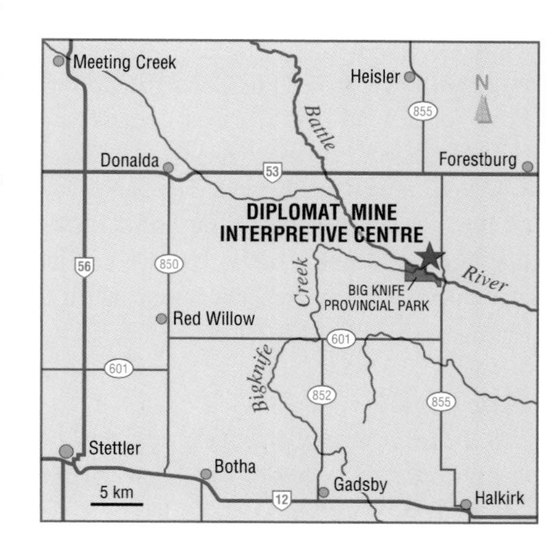

# WOLF CREEK SAND DUNES AND GOLF COURSE

*Ron Mussieux – Provincial Museum of Alberta*

*Well-manicured flower beds near the clubhouse at Wolf Creek Golf Course. The rolling nature of the course reflects the underlying sand dunes.*

## HIGHLIGHTS

Dunes are a beautiful and fascinating geological feature that are more common in Alberta than anywhere else in Canada. The Wolf Creek golf course, on Highway 2 between Lacombe and Ponoka, lies right in the heart of a dune field that some people would consider a vast and empty wasteland. The land was of limited value for ranching and farming, but golf course designers saw it as an opportunity to create a course with a distinct Scottish links style.

## THE STORY

A dune is a mound, hill, or ridge of windblown sand, either bare or variably covered by vegetation. It is capable of blowing from place to place, but retains its own characteristic shape for an extended period of time.

The Wolf Creek dunes lie on what was once the basin of Glacial Lake Red Deer, a large shallow lake formed by the melting of continental glaciers, beginning at least 12,000 years ago. Water flowing from the melting glaciers deposited huge amounts of sand on the bottom of the lake, and as it dried out, a barren, flat area of sand was left behind. This surplus sand and a general lack of vegetation allowed for transverse (at right angles to wind direction) dunes to form. Over time, these dunes changed into "blowout" dunes because vegetation anchored some areas while other sections were blown downwind.

Most of the dunes present at Wolf Creek are blowout dunes. They have an average height of eight metres and are arranged in an overlapping pattern. The wings of the dunes point in a northwest direction indicating that the ancestral wind was not that much different from the predominant wind direction today. If you look closely at the dunes, you can see that many of them have dark, carbon-rich soil bands. These represent former land surfaces that developed during brief periods of increased moisture and decreased winds.

Ron Mussieux – Provincial Museum of Alberta

*Narrow fairways bordered by long grass and abundant, easily constructed sand traps make this a difficult golf course.*

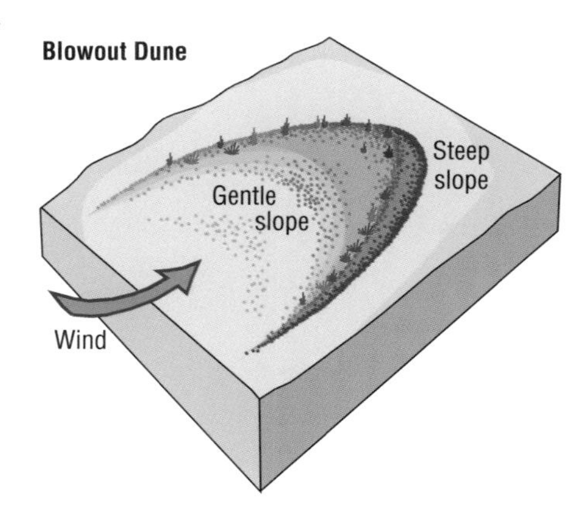


The soil in this area is light and sandy while the inter-dune area is often swampy; thus, this type of land is of little use for agriculture. Also, the sand is too impure to be used for glassmaking. However, the rolling topography, the long grass, and the sand create a picturesque landscape and a challenging golf course. The Wolf Creek golf course is different from most of the courses in this province because it was built on a geological feature that is rarely considered for that purpose. It takes full advantage of the unusual landscape and uses the natural features of the dune field to provide an interesting, naturally contoured golf links.

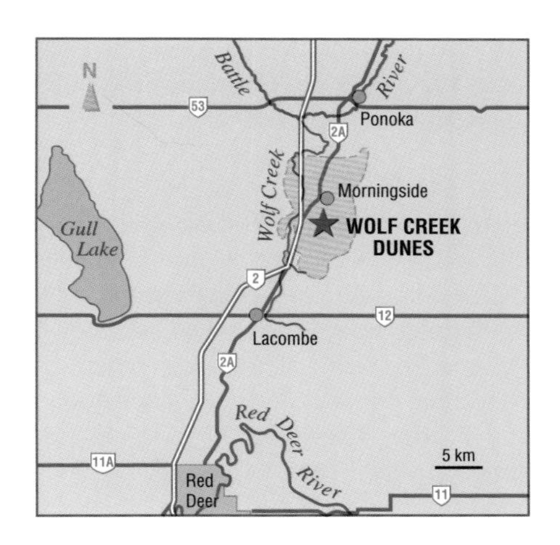

## RAM FALLS: A Classic Example of Differential Erosion

Alberta Resources Service

*The resistant Cardium Formation sandstone separates the underlying and overlying weaker black shales.*

### HIGHLIGHTS

The geology surrounding Ram Falls is spectacular and has been studied by geologists for decades. The combination of folded and faulted rock layers plus the erosive power of water has created this unique landscape. Access to the viewpoint of the falls is immediately off the Ram Falls campsite.

### THE STORY

At this locality, the Ram River cuts through layers of marine sandstone and shales that were deposited about 100 million years ago, during the Cretaceous Period. The falls are formed where a particularly hard layer of sandstone, called the Cardium Formation, is exposed by the flowing water. Sandstone is harder and more durable than shale and therefore erodes more slowly. This uneven erosion rate of rocks is called differential erosion and is responsible for the formation of many waterfalls. Looking upstream from the falls, you can see that the water has easily cut down through the soft layers of shale. When it encountered resistant sandstone however, erosion slowed considerably. The water flowed over and finally cut underneath the sandstone. It then eroded the underlying shale, and gradually a prominent sandstone ledge formed over which the water plummets. The constant pounding of water beneath the falls continues to deepen the plunge pools. As the Cardium sandstone layer extends upstream from the falls but not downstream, the valley floor above is quite flat whereas below the falls it is distinctly V-shaped.

During mountain building times, the rock layers in this area were folded into a small arch, or anticline, which was then shoved over on to its side. Looking across the Ram River to the south bank, and using the Cardium sandstone as a guide, you can follow the upper limb of the anticline east from the falls. The limb, which contains a small fault, runs nearly horizontal and then angles vertically down forming the rapids below the falls.

The Cardium Formation is a thin, but widespread, layer over much of western Alberta and plays an important role in our economic history. It is the main reservoir rock of the enormous Pembina oil field, and it is therefore important for geologists to study where it outcrops at the surface. The Cardium Formation is thought to have been deposited as an offshore sand bar. It is not particularly porous, but because it is overlain and underlain by thick impermeable shales and is wedge-shaped (see page 88), this formation has proved to be a highly effective petroleum trap.

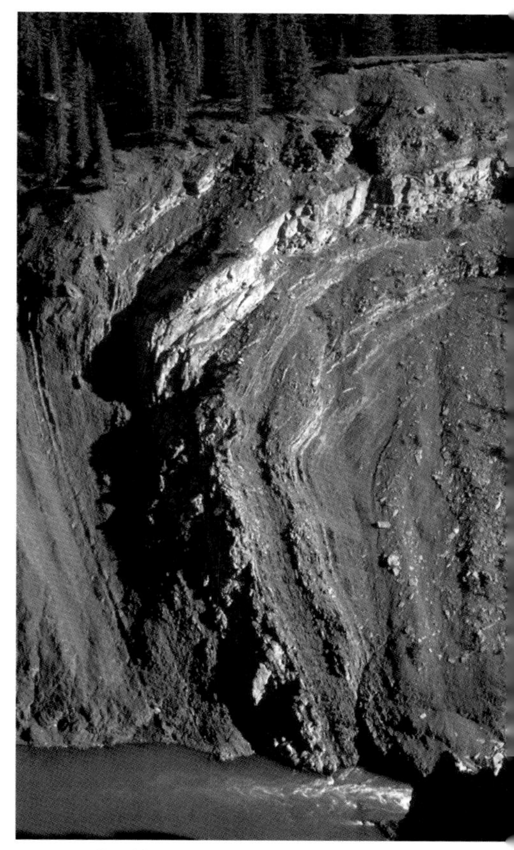

Ron Mussieux – Provincial Museum of Alberta

*A short distance downstream from the falls, the Cardium sandstone curves downwards into an anticline that is lying on its side. Where the sandstone has been eroded by the river it forms a set of rapids.*

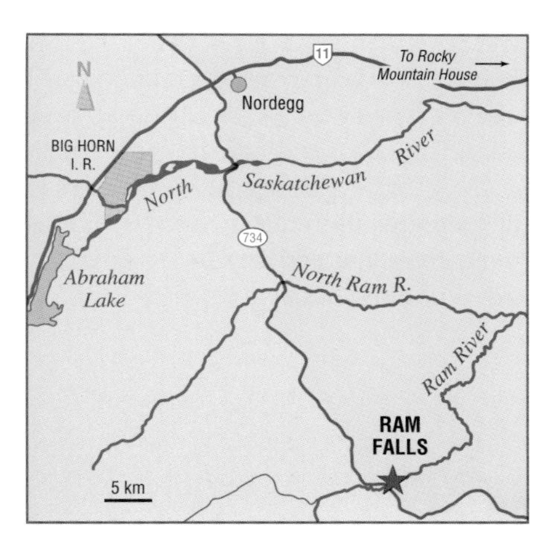

# THE DAVID THOMPSON HIGHWAY:
## A Magnificent Vista of the Rockies

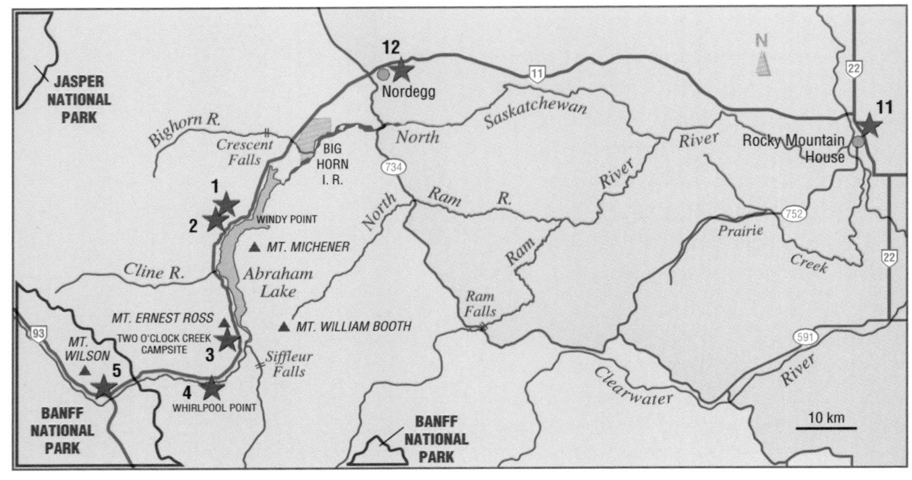


*The David Thompson Highway crosses both the Heartland and Rockies tourist destination regions.*

## HIGHLIGHTS

The David Thompson Highway (Hwy. 11) has been described as one of the ten most beautiful drives in North America. Travelling west from the town of Rocky Mountain House towards Highway 93, the Highway passes from the Plains through three major zones of the Rockies: Foothills, Front Ranges, and Main Ranges. The highway follows the North Saskatchewan River valley which cuts the mountain ridges almost at right angles and allows the visitor an end-on view of several mountain ranges as well as an opportunity to view spectacular geological features such as faults, anticlines, and synclines.

## THE STORY

This description of the David Thompson Highway takes the form of a roadlog and includes several highway stops between Rocky Mountain House and Saskatchewan Crossing. Additional sites worth visiting on this highway, and discussed elsewhere in the book, are Ram Falls, Crescent Falls, Siffleur Falls, and Hummingbird Reef. The Interior Plains, the Foothills, the Front Ranges, and the Main Ranges all have their own characteristic geological properties, such as rock types, age of rocks, attitude of rock beds (horizontal, vertical, or dipping) and type and degree of deformation (folding, faulting).

# An Idealized Cross section Along the David Thompson Highway Showing the 7 Roadlog Stops

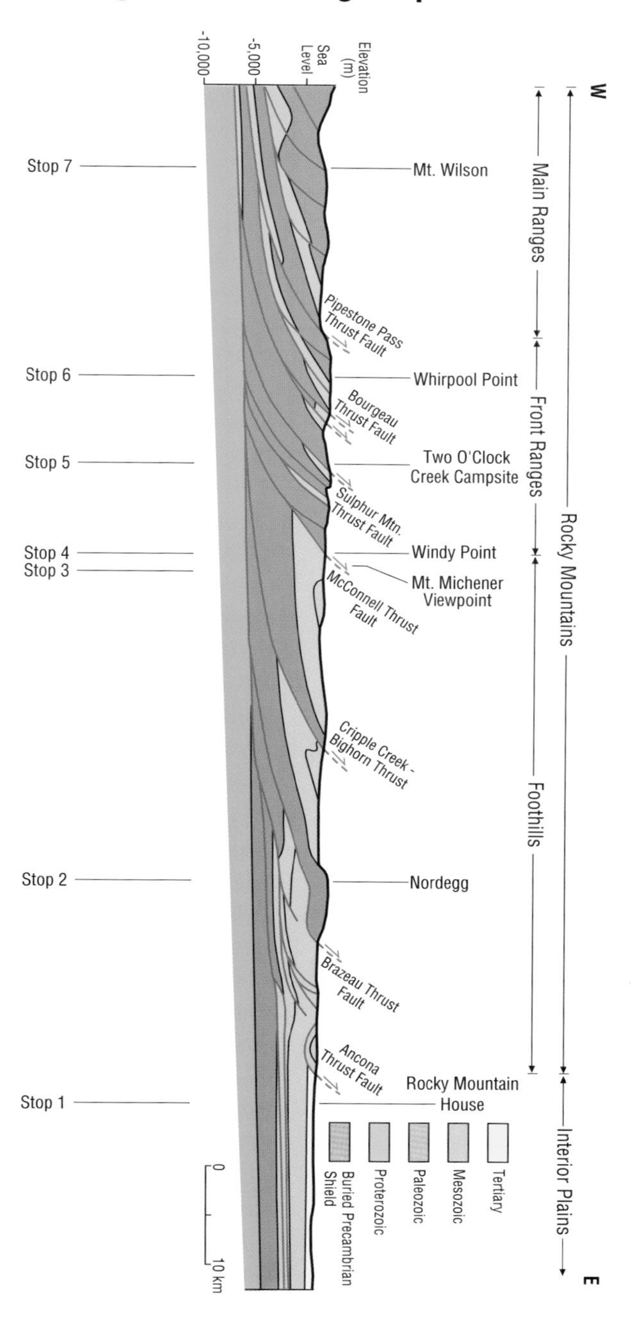

## Regional Characteristics of the Rocky Mountains and the Plains along the David Thompson Highway

| | Rocky Mountains | | | Interior Plains (Rocky Mountain House) |
|---|---|---|---|---|
| | Main Ranges | Front Ranges (Mt. Michener/ Windy Pt.) | Foothills (Nordegg) | |
| Maximum Elevation | Mt. Saskatchewan 3,344m<br>Mt. Wilson 3,192m | Mt. Michener 2,462m<br>Mt. William Booth 2,673m | Mt. Coliseum 2,014m<br>Ram Tower 2,173m | 975 m |
| Age of Rock Units | -mostly Precambrian and Cambrian, some Devonian and Mississippian | Middle Cambrian through to Cretaceous | - primarily Cretaceous inliers of Paleozoic carbonates (Nordegg) | Tertiary |
| Most Common Rock Types | - primarily quartzite, limestone, dolostone | Mountains -primarily limestone and dolostone of Mississippian and Devonian age, some Cambrian<br>Valleys -shale and sandstone of Jurassic and Cretaceous age | Cretaceous -sandstone, shale, coal<br>Paleozoic -limestone (Nordegg) | sandstone |
| Attitude of Rock Beds | - generally flat-lying to gently dipping | - often steeply dipping<br>- frequent tight folds<br>- dip of beds variable | - wide range: steeply dipping to nearly horizontal in the centre of some folds | - nearly horizontal (gentle dip to the west) |
| Rock Structures | - broad open folds<br>- fewer thrust faults | Folds -tight anticlines and synclines<br>Thrust faults<br>Folded faults | Folds -anticlines<br>-synclines<br>Faults -mostly west-dipping | |
| Overall Shape and Appearance of Mountains and Foothills | - gently dipping layered mountains (Mt. Edith Cavell)<br>- broad syncline mountains (Mt. Kerkeslin)<br>- glacially carved horns (Mt. Assiniboine)<br>- flat layer cake mountains (Castle Mountain) | - steeply dipping layered mountains (Mt. Rundle)<br>- complex mountains, extensive folding and faulting (Mt. Michener)<br>- main North Saskatchewan River broad and U-shaped valley (glacially carved)<br>- most tributary valleys are narrow and V-shaped | - narrow, parallel sandstone ridges NW-SE, parallel to main mountain ranges<br>- some folding<br>a) Brazeau Range is a limestone-cored anticline<br>b) Cretaceous sandstone and shale are also folded | |

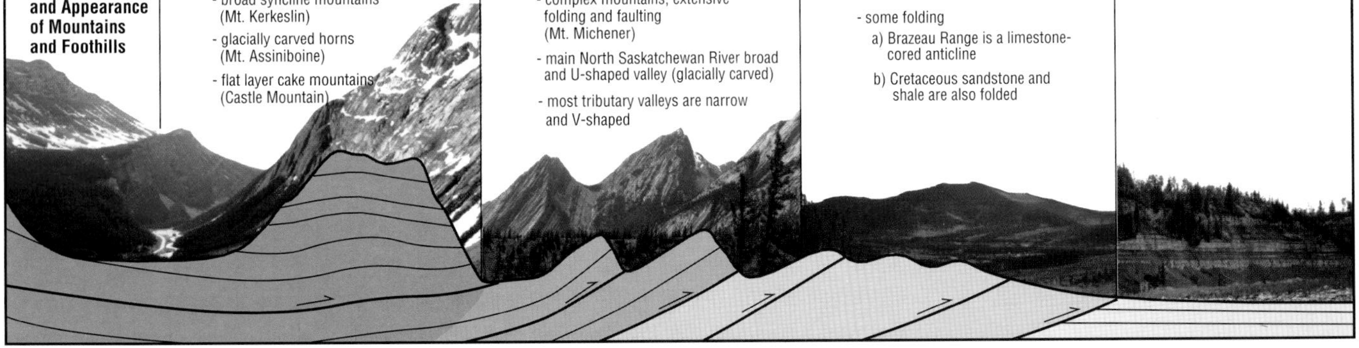

# THE DAVID THOMPSON HIGHWAY: Stop 1

## The Plains: Rocky Mountain House

*Ron Mussieux – Provincial Museum of Alberta*

*Flat-lying sandstone of the Interior Plains exposed along the North Saskatchewan River near Rocky Mountain House.*

The Interior Plains is the relatively flat region that covers the central part of North America. Rocky Mountain House lies near the western edge of the Plains Region of Alberta. Here, the North Saskatchewan River has eroded down through the cliff-forming sandstones of the Paskapoo Formation of Paleocene age, deposited between 65 and 58 million years ago. These rock beds appear horizontal, but they are actually dipping very gently to the west. These tough sandstones are the same rocks that were quarried as building stone in the Pembina River valley (see page 90) and the Calgary-Cochrane quarries (see page 178) and can be seen in many government buildings, schools, and churches across Alberta. Forty-five kilometres west of Rocky Mountain House, near Jackfish Creek, lies the boundary between the Foothills and Plains: the Ancona Thrust Fault. In some roadside outcrops around Jackfish Creek, you will see the first signs of the deformed rocks of the Foothills region.

Of interest near Rocky Mountain

House is a 500-tonne glacial erratic which lies on the south side of Highway 11, just east of its junction with Highway 22. The Rocky Mountain House Erratic and others at nearby Crimson Lake Provincial Park represent an end of the Foothills Erratics Train (see page 176). A group of the Erratics Train has also been mapped north of the McLeod River. No doubt there are erratics between these areas but they are hidden by dense forest.

*Ron Mussieux – Provincial Museum of Alberta*

*A 500-tonne glacial erratic, part of the Foothills Erratics Train, sits by Highway 11 at Rocky Mountain House.*

# THE DAVID THOMPSON HIGHWAY: Stop 2

## The Brazeau Range at Nordegg: Limestone Mountain in the Foothills

*Jane Ross – Provincial Museum of Alberta*

*On Coliseum Mountain (part of the Brazeau Range) at Nordegg, resistant limestone has been folded into an arch shape, called an anticline.*

As you approach Nordegg, the massive Brazeau Range dominates the skyline, extending 55 kilometres in length and rising to an elevation of 2174 metres. Although it looks like you have reached the Front Ranges, they are still another 35 kilometres to the west. The Brazeau Range is actually an oddity in the Foothills — an isolated area of older limestones that is surrounded by the younger rocks of the Foothills. The Range is an anticline of Paleozoic limestones and dolostones that have been greatly uplifted and folded by movement along the Brazeau Thrust Fault. Millions of years of erosion of the softer overlying Cretaceous sedimentary rocks have left the folded limestone standing as a high ridge.

The Brazeau Range is composed of the familiar Palliser and Banff Formations and the Rundle Group which form many other peaks in the Front Ranges. At the Nordegg townsite, if you look northwest along the range towards Coliseum Mountain, there is a gentle arch, or anticline, of these limestones and dolostones which forms the summit of this mountain. Just east of the town, Nordegg Lime Limited is quarrying pure limestone for ornamental rock and as a conditioner for acidic soils.

On the west side of the Brazeau Range is the western limb of the anticline where sandstones, shales, and economic coal seams of the Luscar Group are exposed at the surface. Brazeau Collieries was established in 1911 by Martin Nordegg to mine this bituminous coal as fuel for steam locomotives. The mine produced until 1955 when the railways completed their switch to diesel locomotives. Today, the mine site is designated an Alberta Historic Resource and tours can be arranged through the Nordegg Historical Society.

# CHAPTER 5: THE ROCKIES

*Hiking to Hummingbird reef.*
*Ron Mussieux – Provincial Museum of Alberta*

# ROCKIES

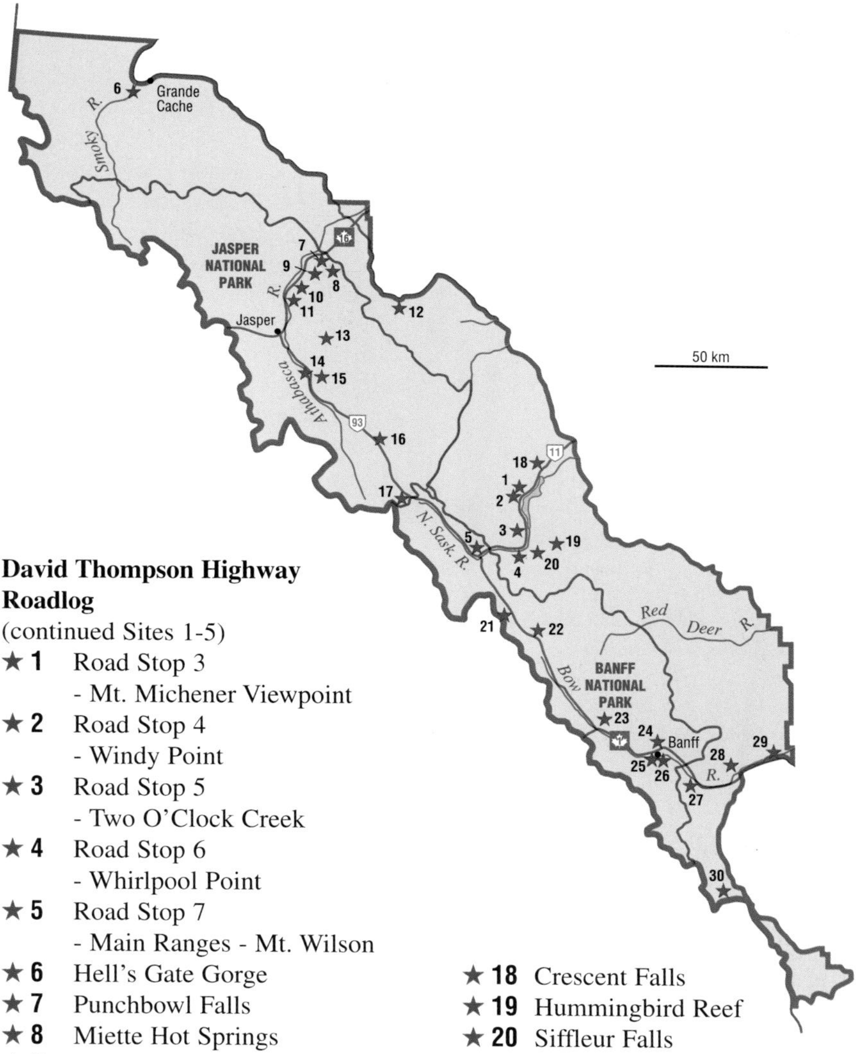

**David Thompson Highway Roadlog**
(continued Sites 1-5)

★ **1** Road Stop 3
- Mt. Michener Viewpoint

★ **2** Road Stop 4
- Windy Point

★ **3** Road Stop 5
- Two O'Clock Creek

★ **4** Road Stop 6
- Whirlpool Point

★ **5** Road Stop 7
- Main Ranges - Mt. Wilson

★ **6** Hell's Gate Gorge
★ **7** Punchbowl Falls
★ **8** Miette Hot Springs
★ **9** Roche Miette
★ **10** Jasper Lake Dunes
★ **11** Cold Sulphur Springs
★ **12** Cadomin Cave
★ **13** Medicine Lake
★ **14** Athabasca Falls
★ **15** Mt. Kerkeslin
★ **16** Jonas Slide
★ **17** Athabasca Glacier
★ **18** Crescent Falls
★ **19** Hummingbird Reef
★ **20** Siffleur Falls
★ **21** Peyto Lake
★ **22** Crowfoot Dyke
★ **23** Castle Mountain
★ **24** Bankhead
★ **25** Mt. Rundle
★ **26** Bow Falls
★ **27** Grassi Lakes Reef
★ **28** Mt. Yamnuska
★ **29** Morley Drumlins

# THE DAVID THOMPSON HIGHWAY: Stop 3

## Mt. Michener Viewpoint: A View of the McConnell Thrust Fault

Ron Mussieux – Provincial Museum of Alberta

*Complex folding of limestone beds on Mt. Michener. Notice the tight anticlines and synclines near the summit outlined by fresh snow. (Above)*

*Deformation in weak sandstones and shales that underlie the McConnell Thrust Fault. Notice the syncline/anticline pair. (Below)*

The Mt. Michener Viewpoint, 33.8 kilometres west of Nordegg on Highway 11, is on the south side of the road overlooking Lake Abraham. This viewpoint is near the boundary between the Foothills to the east and the Front Ranges to the west. The boundary is defined by the McConnell Thrust Fault which can be traced for 400 kilometres along the mountain front. The most distinctive feature of the Front Ranges is the massive, grey wall of Paleozoic limestones and dolostones that tower over the younger and softer Mesozoic sandstones and shales of the Foothills. Remember the Brazeau Range at Nordegg is an oddity — a rare, isolated limestone ridge within the Foothills belt.

The rocks exposed along the road cut north of the viewpoint are composed of sandstones, shales, and coal seams of the Luscar Group which were deposited about 144 million years ago. High above the highway, the summit of First Range is formed of thick, erosion-resistant Cambrian limestone, which is about 510 million years old and tilts to the southwest at a 30° to 40° angle. The trace of the thrust fault which has brought this older rock over the younger rock is hidden by rock debris and vegetation. The powerful forces that moved the extremely tough limestone along the fault plane have buckled the underlying weak sandstone and shales into a complex series of anticlines and synclines, with some minor faulting. This type of deformation that underlies a major fault is often called a "drag fold."

Looking south from the viewpoint, across Lake Abraham, you see two distinctive mountains. The more easterly one is an unnamed mountain that looks very similar to Mt. Rundle in Banff. The McConnell Thrust Fault lies under the east side of this mountain. Rocks as old as Upper Cambrian are exposed near the base of the mountain. The steep cliffs and slopes above these rocks are the Palliser and Banff Formations and the summit is Rundle Group. Further to the west is the larger and more structurally complex Mt. Michener. The north face of Mt. Michener is composed primarily of the Palliser, Banff, and Rundle rocks which are folded into a series of tight anticlines and synclines.

# THE DAVID THOMPSON HIGHWAY: Stop 4

## Windy Point: Evidence of the Work of Ice, Water and Wind

*Strong winds channelled down the river valley whip up silts and clays and deposit them at Windy Point.*

*Jane Ross – Provincial Museum of Alberta*

Two kilometres west of the Mt. Michener Viewpoint, a road leads a few hundred metres south of Highway 11, giving access to Windy Point and another spectacular view of Mt. Michener. Limestones and dolostones of Upper Cambrian age, the oldest rocks exposed along the McConnell Thrust Fault in this area, form a low ridge that projects southward into Lake Abraham. These tough rocks have been smoothed and polished by the abrasive action of an advancing glacier. Larger rock fragments in the base of the glacier have also scratched grooves, called striations, which can be used to determine the direction of ice movement. This bedrock surface was eroded by the advancing Saskatchewan glacier between 22 and 15 thousand years ago. In several areas, these rocks look similar to the recently exposed, polished bedrock at the toe of the Athabasca glacier (see page 144).

Parts of this polished bedrock have been exposed to extensive chemical weathering. Low temperatures, abundant snow melt, a high concentration of carbon dioxide in the snow, and an exposed limestone surface have combined to increase the effectiveness of chemical weathering. Cold water can hold a greater amount of carbon dioxide than warm water and it produces a weak carbonic acid which will dissolve the limestone surface. On a moderate slope like Windy Point, the surface and jointed planes become etched and cut by rills or grooves which are separated by sharp ridges.

Another geological process observed at Windy Point is the transport of small rock particles by strong local winds. Silt and clay particles have been carried from the exposed lake bottom of Lake Abraham and laid down on the limestone as a one-metre thick deposit. Silt and clay deposits that have been transported by wind are called loess.

# THE DAVID THOMPSON HIGHWAY: Stop 5

## The Front Ranges at Kootenay Plains near Two O'Clock Creek Campsite

*Ron Mussieux – Provincial Museum of Alberta*

*Mt. Ernest Ross. Less frequently noticed bright red Triassic rocks overlie grey limestones.*

In the Front Ranges, the steeply dipping thrust sheets, often composed of rocks with widely differing resistance to erosion, lead to more complex mountain structures and more variable mountain shapes than in the Main Ranges. The Front Ranges are a landscape dominated by narrow, steeply dipping mountain ridges and narrow valleys paralleling the trend of the rock beds.

The Two O'Clock Creek campsite is on the Kootenay Plains, 31 kilometres west of Windy Point. Just south of this site the North Saskatchewan River has made a sharp bend and changed from an easterly flow to a northerly flow which now parallels the direction of the mountain ranges. Here, the valley widens considerably and is called the Kootenay Plains. The broad valley is in the rain shadow of the high ranges to the west and thus is warmer and receives less rainfall than the surrounding areas. A grassland vegetation has partially replaced the forest cover and the valley is an important winter pasture for wildlife.

Northward from the campsite area, the west side of the valley is dominated by Mt. Ernest Ross. Highway 11 follows very closely the surface trace of the Sulphur Mountain Thrust Fault along the east side of Mt. Ernest Ross. The main mass of the peak is made up of Palliser, Banff, and Rundle rock units which have been folded into a syncline-anticline pair. The distinctive red rocks making up part of the summit are weathered siltstones. These were deposited during the Triassic Period, between 240 and 210 million years ago on a shallow ocean shelf.

Northeast across the river lies the northward extension of Mt. William Booth which is also formed of Palliser-Banff-Rundle rock units but it was uplifted on a more easterly thrust fault. The river valley has been carved in the weak, easily eroded shales of the Fernie Formation — this is the reason for the rapid increase in the width of the valley. This viewpoint allows you to look directly onto the west face of Mt. William Booth and see the effect of erosion on the steep west-dipping Palliser-Banff-Rundle sandwich. Minor streams have cut through the erosion-resistant Rundle rock and then quickly excavated a V-shaped valley through the Banff Formation until they reach the stronger Palliser rocks. The result is a spectacular series of triangular-shaped facets on the west slope which geologists call "flatirons" because of their resemblance to old-fashioned clothing irons.

*Ron Mussieux – Provincial Museum of Alberta*

*Northern extension of Mt. William Booth. Erosion of steeply dipping limestones and shales form triangular-shaped facets called "flatirons."*

# THE DAVID THOMPSON HIGHWAY: Stop 6

## Whirlpool Point

*Ron Mussieux – Provincial Museum of Alberta*

*Resistant rocks, of Cambrian age, constrict the river flow and form the whirlpool.*

*Ron Mussieux – Provincial Museum of Alberta*

*Fractured black limestone has been filled with veins of white calcite. Outcrop above the highway at Whirlpool Point.*

Whirlpool Point is 36 kilometres west of Windy Point (162 kilometres west of Rocky Mountain House). It gets its name from a long, low ridge of durable dolostone of Middle Cambrian age that forms a major obstruction in the river bed. The river has cut a narrow breach through the ridge but much of the water flow is deflected back upstream by the rock where it forms a large circular eddy, or whirlpool.

Highway 11 crosses the poorly exposed Bourgeau Thrust Fault about two kilometres east of Whirlpool Point. This fault is different from the other faults to the east because it has carried older rocks to a much higher elevation. These rocks, primarily thick, resistant quartzites, dolostones, and limestones of Precambrian and Lower Cambrian age, form most of the Main Ranges further to the west.

Excellent outcrops of these rocks occur in roadcuts on both sides of the highway. The north roadcut directly above Whirlpool Point is a dark grey dolostone of the Middle Cambrian Cathedral Formation which is highly fractured and crisscrossed with white veins of the mineral calcite. Most likely this fracturing occurred during movement along the Bourgeau Thrust Fault. Hot water passing through the fractures dissolved some of the dark dolostone and later deposited white calcite into the cracks. This is the same process that forms valuable sulphide mineral deposits, although little can be found in this outcrop. The Cathedral Formation was mined for lead and zinc at the abandoned Monarch and Kicking Horse mines near Field in Yoho National Park.

# THE DAVID THOMPSON HIGHWAY: Stop 7
## The Main Ranges from Mt. Wilson

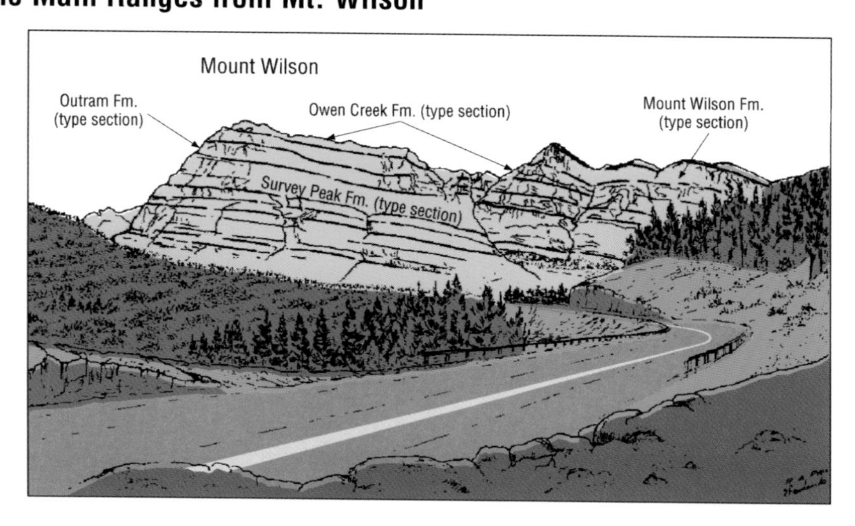


Eight kilometres west from Whirlpool Point, Highway 11 passes over the Main Ranges Thrust Fault, known locally as the Pipestone Pass Thrust Fault. There is some debate among geologists whether this fault or the Bourgeau Fault at Whirlpool Point forms the eastern boundary of the Main Ranges. Within the Main Ranges the mountains are mostly made up of rocks of Cambrian to Ordovician age, and within some of the deeply eroded valleys there are exposed rocks of late Precambrian age. All of the younger, overlying rocks have been removed by erosion. The Main Ranges are less complex than the Front Ranges and are composed of thick, erosion-resistant quartzites, limestones, dolostones, and shales that have been uplifted high into the air. Generally, Main Ranges have fewer thrust faults and the thick, tough rocks have resisted deformation and formed only broad folds with gently tilting beds.

Along the North Saskatchewan River, the higher Main Ranges were more strongly glaciated than the Front Ranges and many of the lofty summits still have active glaciers. Main Range valleys are broad and U-shaped and the mountains are often cut into large isolated masses as opposed to the long, parallel linear ridges of the Front Ranges.

Near Mt. Wilson, at the junction of Highway 11 and the Icefields Parkway (Highway 93), the wide valley of the North Saskatchewan River allows several vistas of many Main Range mountains. Mt. Wilson consists of gently eastward-dipping rocks of Middle Cambrian through to Upper Ordovician ages. Mt. Wilson is a important mountain to geologists because four of its rock formations are called "type" sections. A type section is the standard reference section used worldwide for a particular rock formation. It is rare that one mountain has this many type sections and perhaps the most easily recognized of them is the cliff-forming sandstones of the Mt. Wilson Formation which form the summit of the mountain.

# HELL'S GATE GORGE

*Ron Mussieux – Provincial Museum of Alberta*

*View up the gorge of the Sulphur River. The vertical cliffs on the right side of the gorge are composed of the hard conglomerate of the Cadomin Formation.*

## HIGHLIGHTS

A spectacular 60-metre deep, narrow gorge displaying one of the most prominent rock formations in the Alberta Foothills — the Cadomin Formation — is located 13 kilometres southwest of the town of Grande Cache. Hell's Gate, or Sulphur Gates, as the gorge is known, is at the junction of the Smoky and Sulphur rivers. To reach the day-use area at Hell's Gate, travel west from Grande Cache on Highway 40 for 5.7 kilometres and then turn left on Hell's Gate Road and go another 7.5 kilometres. From here, a short trail takes you to a scenic view of the gorge, where rising from the valley floor are impressive cliffs of the Cadomin Formation.

## THE STORY

The Cadomin Formation is a massive conglomerate, similar in appearance to concrete, that contains pebbles and sand that are tightly cemented together, making this rock hard and resistant to erosion. It was deposited around 118 million years ago in rivers that were flowing east, away from the newly rising Rocky Mountains. As mountain building continued, tremendous forces buckled the rock beds farther and farther to the east until the Cadomin conglomerates themselves were involved. At this location, you can see that through folding and faulting, the conglomerate now stands in a near-vertical position and averages about 45 metres thick. The Cadomin Formation is the same forma-

Ron Mussieux – Provincial Museum of Alberta
*Slow mixing of the waters of the Smoky River (left) and Sulphur River (right).*

tion that outcrops at Kakwa Falls, Punchbowl Falls and also caps many ridges in the western Foothills.

Looking across the Smoky River and up the Sulphur River, you can see the resistant conglomerate that makes up the vertical valley walls of this canyon. Along the trail, small areas of rock are often grooved and polished. This feature, called a slickenside, is evidence of faulting, where two rock masses slid against and past each other leaving a polished surface.

Besides the outstanding rock exposures, the meeting of the two rivers themselves is noteworthy. The rivers have different sources and therefore look quite different. The Smoky River is fed by a glacier and carries a lot of fine-grained sediments that make it quite cloudy. The Sulphur River, however, is fed by rainfall; so it is quite clear. The two waters do not mix right away and you can see water of two different colors flowing side by side for a considerable distance. The Sulphur River forms the northern boundary of the Willmore Wilderness Park, the largest provincial mountain wilderness area.

Ron Mussieux – Provincial Museum of Alberta
*Cliffs of Cadomin Formation.*

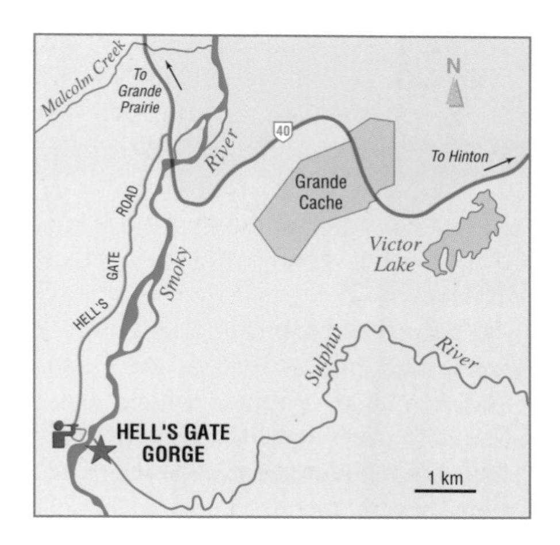

## PUNCHBOWL FALLS OF MOUNTAIN CREEK

Jasper National Park

*Series of small falls and plunge pools in Cadomin Formation.*

### HIGHLIGHTS

At this site, Mountain Creek cuts a narrow cleft into a resistant steep conglomerate cliff, through which water plummets forming Punchbowl Falls. The continual pounding of the water has created several picturesque plunge pools below, giving the falls its name. A short trail to the viewpoint can be reached 1.4 kilometres up Miette Hot Springs Road. The proximity of these falls to Miette Hot Springs and the deserted Pocohontas coal mine makes this area a geologically interesting locality to visit.

### THE STORY

The conglomerate responsible for this waterfall is the Lower Cretaceous Cadomin Formation, which was deposited approximately 118 million years ago. Conglomerate is a sedimentary rock that contains pebbles, cobbles, and boulders and is sometimes referred to as "nature's concrete." Each pebble is a fragment of some older rock and was rounded by tumbling in a fast moving stream before coming to a final resting place. The size of the pebbles and cobbles tells the geologist that we are not a great distance from their source because of the current and steep slopes necessary to move rocks of this size. Rounding suggests a great deal of shifting and reworking in a sediment clogged channel.

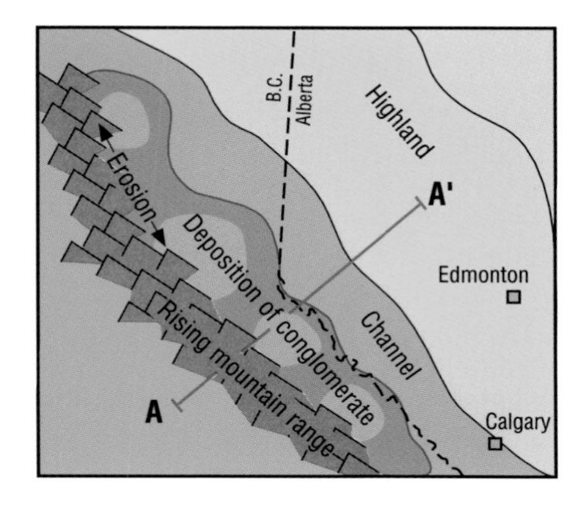


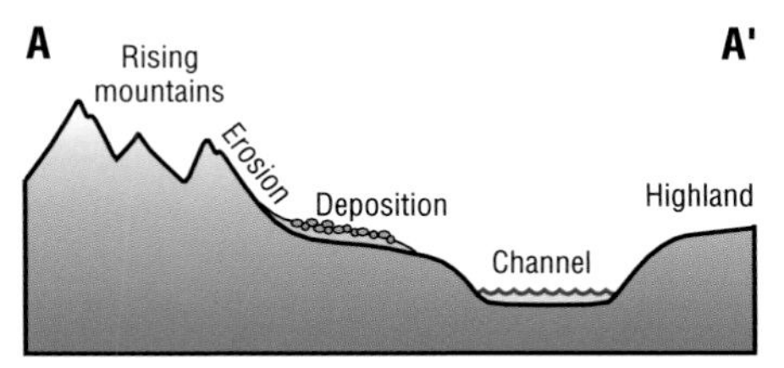


The pebbles in the conglomerate at Punchbowl Falls came from mountains to the west. These mountains were continuously eroded and reduced in height as rivers carried pebbles away and deposited them as a thin blanket on a broad plain. Over millions of years, the pebbles were covered by other sediments and eventually were cemented together. The Cadomin Formation is a hard, extremely durable conglomerate that extends over much of the length of the southern Rockies and Foothills. Continued uplifting since the Lower Cretaceous Period has formed the mountains and the foothills and shoved the conglomerate up into the spectacular cliffs we find in areas such as Hell's Gate near Grande Cache.

Near Punchbowl Falls is the sealed entrance to the Pocohontas coal mine, abandoned in 1921. Crossing Mountain Creek was an everyday occurrence for the miners and they used a decorative bridge built over the falls, where the upper viewpoint is today. There are six coal seams above the Cadomin Formation. The mine was mainly working a single seam 3.6 metres thick. A small seam of coal, about 25 centimetres thick, is visible near the lower viewpoint. Looking across the valley, you can see that the erosion-resistant limestones and dolostones form the ridges, while the softer coal-bearing Mesozoic rocks form the valleys.

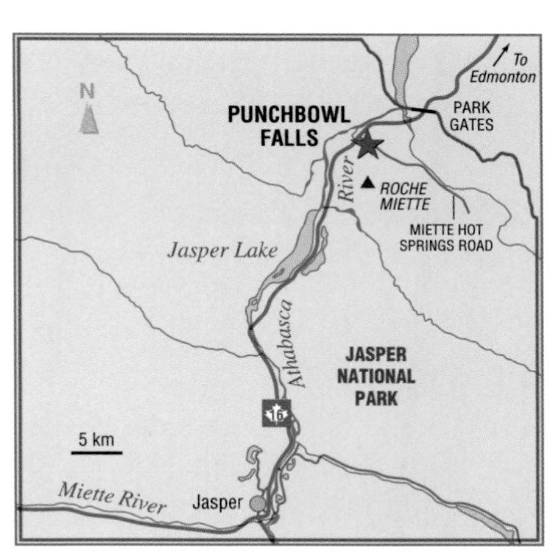

# MIETTE HOT SPRINGS: Nature's Hot Tub

*Marilyn Nelson – Provincial Museum of Alberta*

*The "hottest" hot springs, at almost 54°C, make the Miette Hot Springs pool a major Jasper attraction.*

## HIGHLIGHTS

Miette Hot Springs have the highest measured temperatures in Alberta, at 53.9°C. Hot mineral waters have had a reputation for centuries for their therapeutic and curative powers, and for this reason Miette Hot Springs is a major attraction in Jasper National Park. From a geological perspective, hot springs provide evidence of the faulted and fractured rock beneath our feet, as well as the hot interior of our planet.

## THE STORY

Rainwater slowly filters down through faults, fractures, and pores in rocks. This is the water that later returns to the surface as hot springs. As the water seeps down, it becomes increasingly heated by the surrounding rocks — the rocks are heated by the radioactive decay of some minerals within the earth's crust. In the Rockies, the ground temperature increases about 1°C every 33 metres in depth. After the water has descended several kilometres, it will be heated above its boiling temperature, although the immense pressure beneath the earth's surface will prevent it from actually boiling. Hot water rises upwards and the numerous fractures and faults in the Rocky Mountains provide the perfect conduits for rapid flow back to the surface. The more direct the route, the hotter the water will be when it reaches the surface.

Immediately under Miette Hot Springs, the rock layers have been folded upward into an arch, called an anticline. A nearby fault, imaginatively

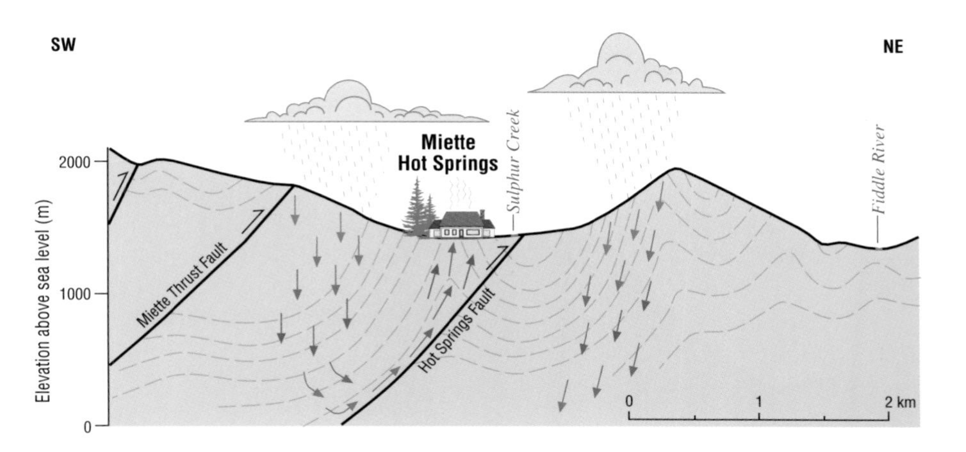


named Hot Springs Fault, likely provided the access route for the hot water to move upwards and into the fractures in the anticline. The water then travels the rest of the way up, via the fractures, and into the Sulphur Creek valley.

Relative to other springs in the Rockies, the water of Miette Hot Springs has a high temperature and a high concentration of dissolved minerals, reflecting a long, deep circulation route. Estimates indicate the water travels upwards from at least three kilometres below, giving it adequate time to dissolve minerals from the surrounding rock. Contact with radioactive minerals occurring at such depths makes hot spring water slightly radioactive.

The crumbling, porous, cream-colored masses around the spring outlets are a type of limestone, called tufa, the youngest rock in the Rockies. As hot water moves up towards the surface, calcium carbonate ($CaCO_3$) is dissolved out of nearby limestone. When the water trickles out, dissolved carbon dioxide gas escapes, and $CaCO_3$ is deposited in layers around the outlet. Tufa deposits can also be found farther upslope from Miette Hot Springs, indicating the springs were once at a higher elevation. Of historical interest, the first development of Miette Hot Springs was initiated by miners of the Pocahontas Mine in 1910, during an extended strike.

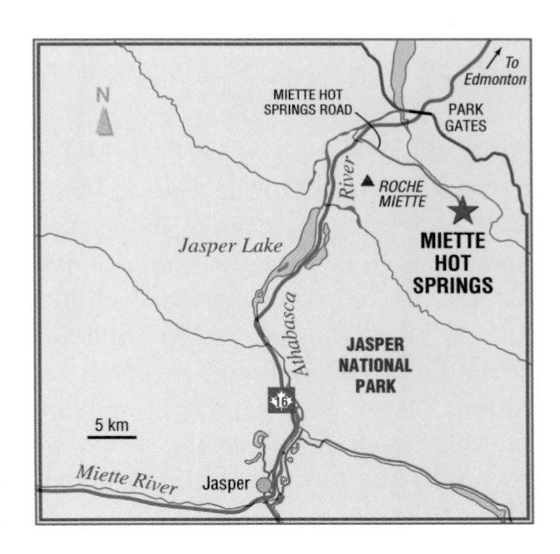

# THE IMPOSING CLIFFS OF ROCHE MIETTE

*Marilyn Nelson – Provincial Museum of Alberta*

*Steep cliffs of Devonian Palliser Formation form the summit of Roche Miette.*

## HIGHLIGHTS

Travelling west from Hinton, you will see that the rocks forming the mountains become increasingly deformed as you go from the Foothills into the Front Ranges. Roche Miette is a well known Front Range mountain in eastern Jasper National Park, and can even be seen from as far away as Hinton. It was named after Miette, a fur trader who climbed to the summit and dangled his legs over the vertical north face. The rocks in this area have been studied intensively by geologists because their subsurface equivalents are reservoirs for much of Alberta's petroleum resources.

## THE STORY

Roche Miette provides an excellent section of spectacularly folded Cambrian and Devonian rocks. The rocks of the main peak were laid down on a tropical ocean floor during the Late Devonian, between 373 and 366 million years ago. The hundreds of metres of grey rock that make up the vertical cliff at the top of the mountain is the Devonian Palliser Formation, a massive and tough limestone that forms many of the steep cliffs seen throughout the Rockies. Along this stretch of highway, Roche Ronde, Roche à Perdrix and Ashlar Ridge are all capped with this same erosion-resistant formation. The

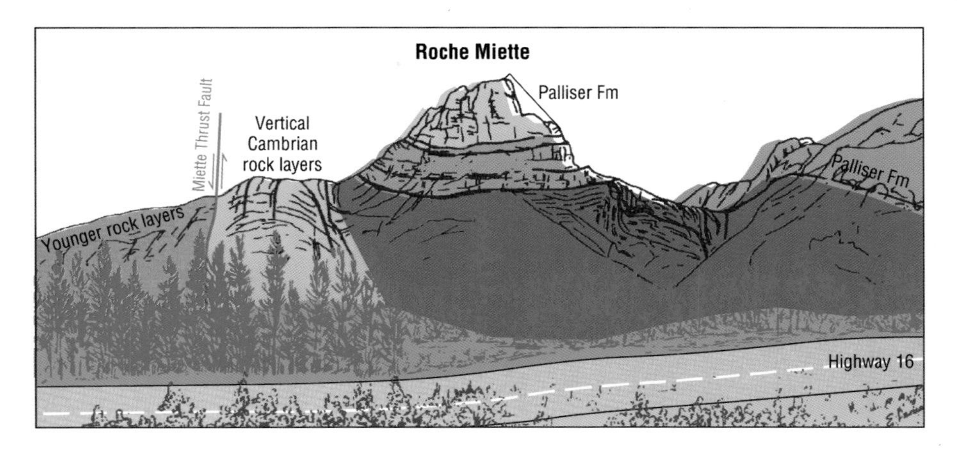


Palliser Formation, which is often compared to the massive limestone banks of the modern Bahama Islands, is important in Alberta because it is the main rock quarried for the cement industry. Below the Palliser Formation are shales that are easily eroded and deformed. These form the crumbling cliffs and tree-covered slopes that make up the base of Roche Miette.

During the time that all of these Devonian sediments were being deposited, southwestern Alberta was covered with a shallow tropical ocean that was dotted with stromatoporoid reefs. One of these reefs, the Miette Reef complex, is just south of Miette Hot Springs. An interpretive panel at the pool discusses the formation of this reef. When Miette reef was growing and flourishing, the sediments that make up much of Roche Miette were accumulating just off the reef.

On the ridge to the left of the main peak are near-vertical Cambrian beds that lie on the steeply dipping Miette Thrust Fault. These rocks contain trilobites, an group of extinct arthropods similar to crustaceans, that once inhabited the seas. Geologists use trilobites to date these rocks at about 520 million years old, making them some of the oldest rocks in the eastern half of the park. Salt crystals, salt casts, ripple marks, and mud cracks are also common in these Cambrian rocks, indicating a shallow salty sea that in places was beginning to dry up.

Thus, the twisted and folded rock layers of Roche Miette provide us with both physical and organic evidence that a once-horizontal ocean floor was lifted up thousands of metres to its present location in the mountains.

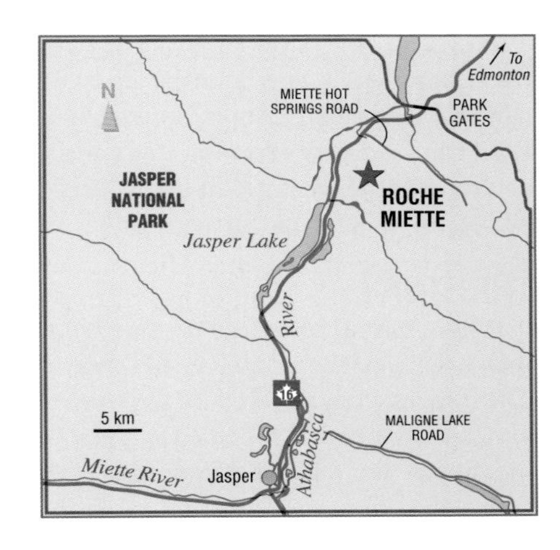

## JASPER LAKE DUNES: Dunes in the Mountains

*Marilyn Nelson – Provincial Museum of Alberta*

*Sand dunes forming along Jasper Lake with Roche Miette in background.*

## HIGHLIGHTS

Alberta contains half of the dune area in Canada. A sand dune is a mound or ridge of sand that is blown into a characteristic shape by wind. Visitors entering Jasper National Park along the Yellowhead Highway (Highway 16) are often surprised to see extensive areas of shifting sand dunes along Jasper Lake. How can sand dunes form in a mountain environment? Basically, all that is needed is a lot of sand and strong winds.

## THE STORY

Where did all of the sand come from that produces the Jasper Lake dunes? The process began over 10,000 years ago, near the end of the last Ice Age. At the site of the current Jasper Lake, an enormous lake was forming as huge glaciers melted and torrents of water poured into the valley. This prehistoric lake stretched for a length of almost 100 kilometres. Lake deposits and terraces now found in the hills above Jasper Lake reveal that the surface of the glacial lake was 100 metres higher than that of the present lake. Over time, the ancient lake drained and sands and silts on the lake bottom were exposed to strong winds and blown into fields of dunes. Vegetation had not yet become firmly established after the melting of the ice and as a result the valley floor became a zone of blowing sands and advancing dunes.

The process of depositing sand at Jasper Lake continues today. Jasper Lake is essentially a wide, shallow portion of the Athabasca River, which is

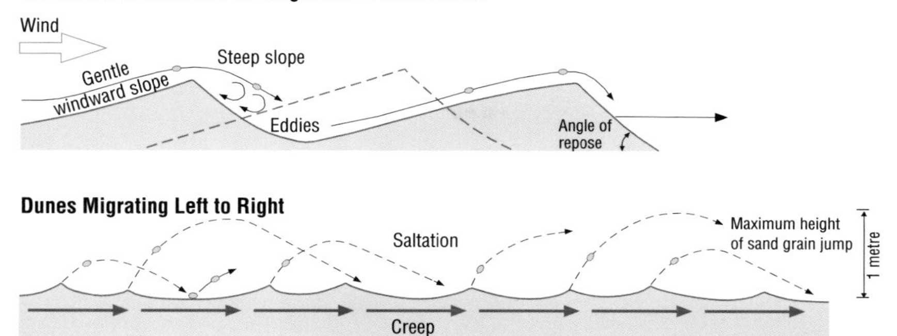


fed by melting mountain glaciers. In the spring, as the glaciers melt, the water level of Athabasca River, and therefore Jasper Lake, goes up. Large amounts of sand and silt from the melting ice are transported and deposited on the shoreline of the lake. During the winter months when the water level is down, the summer deposits of sand and silt are dried by the westerly winds and blown away in spectacular dust storms. The lighter silt stays in the air for a long time and is blown east to be deposited on the Foothills as a thin layer of silt, called loess. The heavier sand grains are blown into the shifting dunes surrounding the lake. Some of the dunes have dammed portions of the main lake leading to the formation of Edna and Talbot Lakes.

*Marilyn Nelson – Provincial Museum of Alberta*

*Sand ripples form on a major dune, Jasper Lake.*

Many of the dunes reach a height of 30 metres. They contain smooth and well-polished stones that have been naturally sand-blasted by smaller wind-blown grains.

# JASPER'S COLD SULPHUR SPRING

*Ron Mussieux – Provincial Museum of Alberta*

*Cold Sulphur Spring emerges from limestone bedrock.*

## HIGHLIGHTS

At this location, spring water with a strong odor pours from the base of a steeply tilted limestone and shale cliff along the southeast side of the highway. The water in this spring is cold, and varies between 7.5° and 9°C, depending on the season. The name is derived from the rotten-egg odor of hydrogen sulphide gas which escapes from the water. Fossils of stromatoporoids and brachiopods can be found in the cliff walls surrounding the spring.

## THE STORY

Like most springs in the Canadian Rockies, Cold Sulphur Spring is associated with a major fault. This fault is exposed near the peak of Hawk Mountain which is southeast of this spring. Geologists believe that the spring is fed by precipitation that falls on Hawk Mountain, trickles down the fault plane, and eventually emerges as spring water at the base of the cliff where the fault is once again exposed. Some of the water likely continues down the fault to unknown depths.

As water seeps down below the earth's surface, it warms and dissolves minerals from the surrounding rocks. Compared to other springs in the Rockies, however, Cold Sulphur Spring carries a lower concentration of dissolved minerals and is quite cold. Thus, its water likely does not travel long distances nor to great depths and any warm water that returns to the surface is diluted by cold groundwater. This spring does, however, contain a high concentration of salt and sulphur.

Pyrite ($FeS_2$), a sulphur-rich mineral that is often called fool's gold, is common in small amounts throughout the Canadian Rockies. In the presence of hot water, sulphide-oxidizing bacteria will dissolve pyrite producing the hydrogen sulphide gas that we smell at many springs. Some of the gas remains dissolved in the water, and some reacts with calcium in the water to form microscopic sulphate minerals which give this spring its milky appearance.

The salt in the water is deposited near the spring outlet and, as a salt lick, is a major attraction for wildlife, much to the enjoyment of visitors and the chagrin of truck-drivers. This high salt concentration is more difficult to understand. Possibly, because most of the Rockies are composed of marine sediments, the seeping water is slowly dissolving a small buried salt bed.

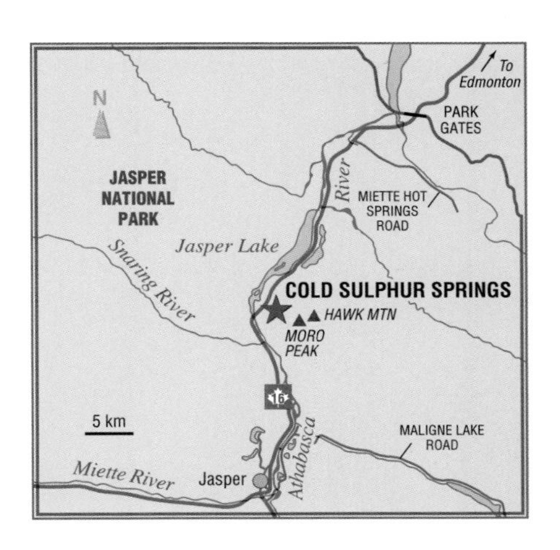

*Ron Mussieux – Provincial Museum of Alberta*

*Spring water deposits white sulphate minerals on rocks and twigs.*

# CADOMIN CAVE: Alberta's Most Accessible Cave

PA.2792/3 – Provincial Archives of Alberta

*A group of cave explorers moves through some of the wider passages in the Cadomin cave in the late 1950s.*

## HIGHLIGHTS

Caves are complex and poorly understood geological features that have long possessed a mysterious aura. Cadomin Cave, located in the mountains above the town of Cadomin, has intrigued geologists and curious tourists for years. The cave is the best known and most accessible one in Alberta, and from the highway it can be seen in the northern slope of Leyland Mountain. So far, exploration of its interconnected rooms and passages has its dimensions at 2791 metres long and 220 metres deep! It is expected that with more exploration, even more passages will be discovered.

## THE STORY

Cadomin Cave is three kilometres south of Cadomin and 366 metres above the McLeod River. As with most caves in the Canadian Rockies, it is formed in limestone — in this case the Devonian Palliser Formation — a tough, hard limestone that forms many steep cliffs in the Rocky Mountains. Caves are formed as acid-rich groundwater seeps through fractures in the bedrock and slowly dissolves the limestone, creating a maze of cavities, passages, and pockets beneath the ground surface. There is only one known entrance to the cave but prospectors have started many rumors of other

Ron Mussieux – Provincial Museum of Alberta

*Series of benches and roads at the Inland Cement Quarry on Cadomin Mountain (1982).*

entrances. It is entirely possible that there are other entrances but they were covered by glacial deposits during the last Ice Age.

The Upper Section of the cave is composed of large passages and rooms with names like the Mess Hall and the Main Gallery. This is the part of the cave that is often visited by tourists and where guided tours are given. Unfortunately though, this human visitation has damaged the cave. The walls are spray painted and there are few rock formations left. Most have been broken off and taken by visitors. The lower and less visited portions of the cave are not as extensively damaged, and they house some interesting cave structures such as stalactites and stalagmites. This cave is also the home and winter hibernation location for about 1000 Little Brown Bats (*Myotis lucifugus*).

Palliser limestone contains beds of almost pure calcium carbonate with few clay or magnesium impurities, making it highly desirable for the manufacture of cement. Across the river valley from Cadomin, the Inland Cement Company is mining this limestone from a large quarry.

***Exploring caves is inherently dangerous and could result in serious injury or death! Under no circumstances should cave exploration be undertaken without proper training, supplies, equipment, and experienced guides. For more information about visiting Cadomin Cave, contact the Alberta Speleological Society or the tourist information centre in Hinton.***

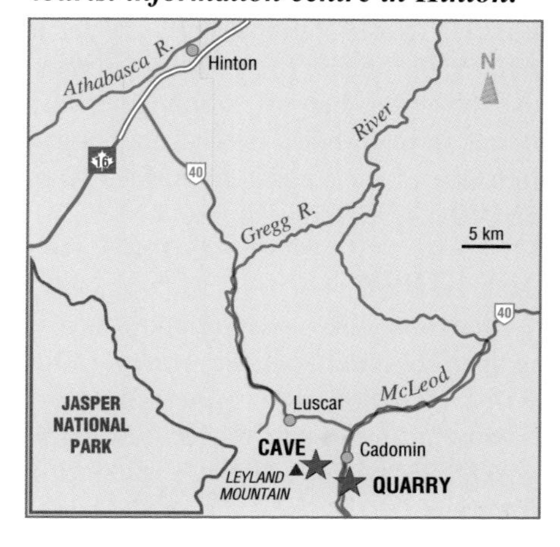

# MEDICINE LAKE: The Disappearing Lake

*Marilyn Nelson – Provincial Museum of Alberta*

*Medicine Lake in the fall when the inflow of water cannot keep up with the subsurface drainage into a cave system.*

## HIGHLIGHTS

An extensive system of interconnected caves and channels beneath Medicine Lake, in Jasper National Park, creates a remarkable annual cycle. In the summer the lake fills, but by autumn the lake drains through holes in the lake bottom into one of the largest subsurface river and cave systems in the world.

## THE STORY

Medicine Lake was formed when a major rockslide from the nearby Colin Range tumbled into Maligne Valley, damming the flow of much of Maligne River. Subsequently, the only outlets for the water became holes in the lake bottom, called sinkholes, located along the northwest and northeast shores. The water now drains through these into subsurface river channels at an incredible rate of 24,000 litres per second! Depending on the season, Medicine Lake has the reputation of being either one of the largest or smallest lakes in Jasper National Park. In the spring, runoff exceeds the drainage capacity of the sinkholes and the lake fills to a depth of 18 metres. Some water may spill over the rock dam at the end of the lake and join the lower Maligne River. By autumn, however, more water drains out through the holes than flows into the lake and Medicine Lake slowly dwin-

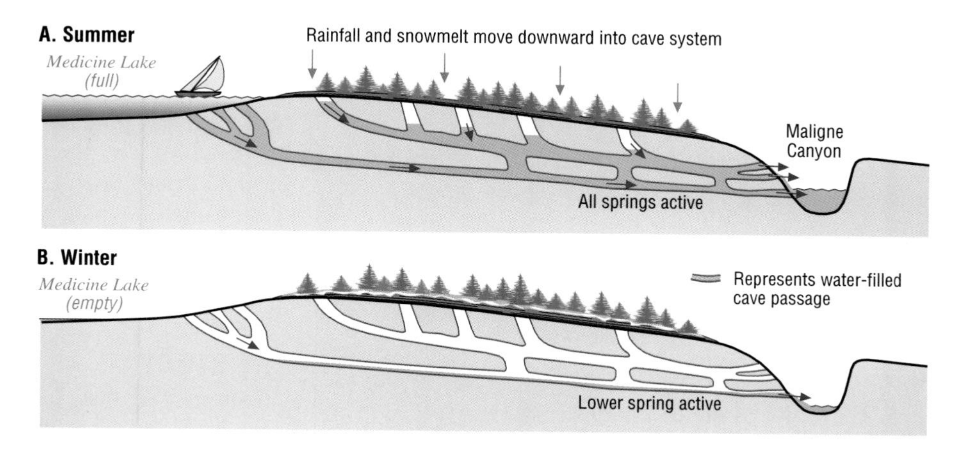


dles. All that is left is a tiny stream trickling across the mudflat. This dry lake bottom is an excellent spot to watch wildlife as it wanders across in search of water.

Where does all the water go? Dye poured into sinkholes shows that the water re-emerges through several springs 17 kilometres downstream in the Maligne Canyon (below the fourth bridge). In the summer it takes only 20 hours for the dye to flow from Medicine Lake to the canyon. In the winter, it can take up to nine days. Whatever the season, this subsurface addition of lake water halfway into the canyon means that more than twice as much water leaves the canyon as originally flowed in!

Medicine Lake and much of the Maligne Valley are underlain by limestone. Groundwater is slightly acidic and as it seeps down through cracks and joints in the valley floor, it dissolves the limestone, slowly creating a vast system of caves and channels. Landforms, such as sinkholes and caves, that are produced when limestone (and less commonly gypsum) is dissolved, are called karst topography. Eventually, the karst beneath Medicine Lake will be enlarged enough to drain all the water year-round, and this lake will cease to exist. Some geologists believe that the cave network beneath Maligne Valley is the longest in Canada. So far, most of it is still unexplored.

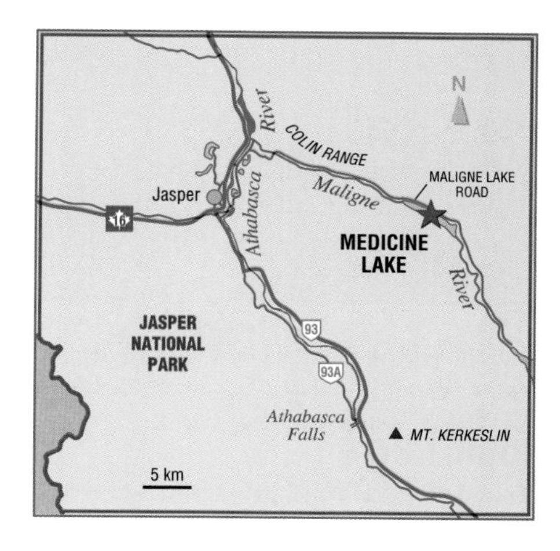

# ATHABASCA FALLS: A Glacial Rock Step

Marilyn Nelson – Provincial Museum of Alberta

*The Athabasca River plunges over an outcrop of Gog Group quartzite, one of the toughest rocks found in the Rockies.*

## HIGHLIGHTS

In a cloud of mist, the Athabasca River plunges 30 metres over a rock step to create this dramatic waterfall. Below the falls, the entire volume of the river is then funnelled through a narrow, steep-walled and dangerous gorge. The combination of glacial erosion, water, and a resistant rock layer has produced this waterfall. The developed trail and interpretive plaques combine to make this a popular tourist attraction along the Icefields Parkway.

## THE STORY

During the last Ice Age, the Athabasca Valley was filled by glaciers grinding northwards from the Columbia Icefield. As they advanced, they carved the valley into a smooth, U-shaped trough. The extremely hard quartzite of the valley floor was able to resist glacial erosion however, and as the flowing ice plucked blocks of quartzite from the bedrock, a stepped surface was gradually created. The ice continued to excavate the step and slowly carved it back and downward. Today, this step is the ledge over which the river water cascades to form the Athabasca Falls.

Since the retreat of the glaciers, the Athabasca River continues to erode the underlying quartzite step, moving the ledge upstream a few millimetres each year. This pale grey quartzite is about 570 million years old and geologists call

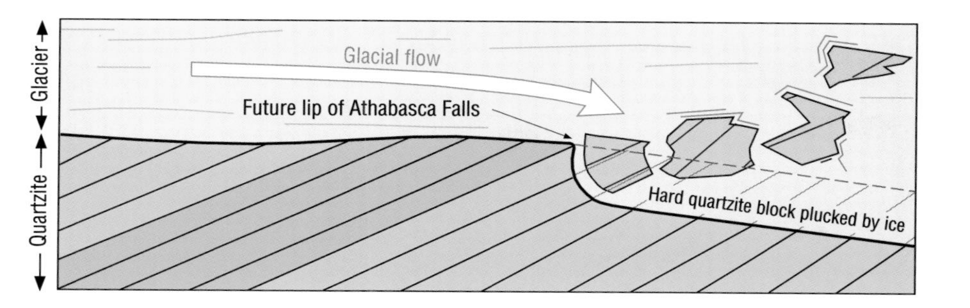


it the Gog Group. It is the hardest rock in the Rockies and tends to form broad waterfalls with short canyons. Other waterfalls in the Rockies are underlain by sandstone or limestone, both more easily eroded than quartzite, and are narrow and high with deep, long canyons.

Many geologists believe that a glacier forced the Athabasca River from its original valley, over a divide, and into the modern Athabasca Valley. Near the falls, potholes and abandoned river channels tell a story of a river that has changed its course many times — a long history of flowing water searching for the path of least resistance. Potholes, created by the scouring action of grit or sand in water, can be found downstream of the falls. Look for old river channels along the west bank.

Fine clay-sized sediments in the water give the river a milky appearance. These sediments, called rock flour, are produced where rock has been ground up beneath the glaciers that feed the Athabasca River.

From Athabasca Falls there is an excellent view of Mt. Kerkeslin. You can see that the rock layers of this mountain have been folded into a U-shape, which geologists call a syncline.

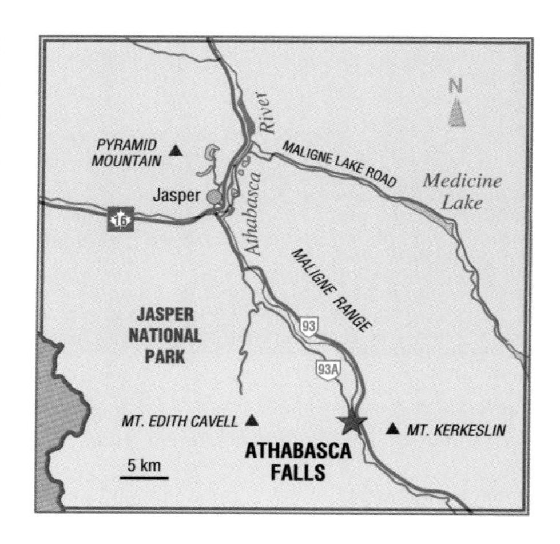

# MOUNT KERKESLIN: A Synclinal Mountain

*Edgar T. Jones – Provincial Museum of Alberta*

*Broad, gentle downward folds, called synclines, are common features of the Main Ranges. Compare the syncline of Mt. Kerkeslin to those of Mt. Michener.*

## HIGHLIGHTS

Compared with mountains in the Front Ranges, Main Range mountains, such as Mt. Kerkeslin, are generally made up of older rocks, usually of late Precambrian, Cambrian, and Ordovician ages. Extensive glacial erosion of the main valleys and their tributaries has left many of these mountains as isolated masses. Unlike the Front Ranges, the rocks in the Main Ranges have been deformed into broad, gentle, open folds. One type of fold, where rock layers bend downwards from each side towards a central axis, is called a syncline. Mt. Kerkeslin is a classic example of a synclinal mountain. The syncline can be seen particularly well from Athabasca Falls.

## THE STORY

When the rocks that make up Mt. Kerkeslin were deep beneath the earth's surface, they were warm and flexible. During mountain building, between 180 and 50 million years ago, these sedimentary rocks were compressed, warped into a variety of folds, and uplifted to form part of the Rocky Mountains. Folding is seen on many scales, from a small outcrop of rock next to the road to an entire mountain.

Synclines are usually paired with arch-shaped folds called anticlines. One might expect synclines to form valleys and anticlines to form mountains, but sometimes the opposite occurs. Rock layers of synclines were compressed and compacted during folding and are

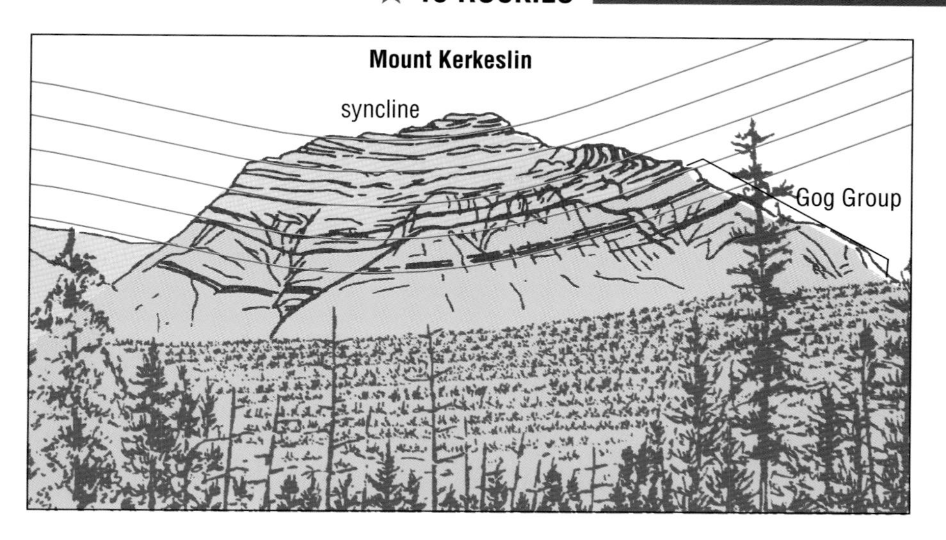


quite difficult to erode. The rocks of anticlines, however, were stretched and fractured as they bent. The fractures provided ice and water the opportunity to wear away the rocks and often valleys, such as the Athabasca and Mistaya river valleys, are formed where anticlines once existed.

Mt. Kerkeslin marks the northern end of an ancient 260-kilometre-long syncline that once extended northwest from Castle Mountain. Erosion over millions of years has reduced this long syncline to the many peaks and valleys we see today. At Sunwapta Pass, the view southeast down the valley reveals mountains as far south as Banff National Park that were once part of this long structure. Nigel Peak to the northeast is another example.

Mt. Kerkeslin is made up of rocks that were deposited between 545 and 510 million years ago on an ancient ocean floor. The two red layers immediately above the tree-line contain iron oxide, called hematite, which gives them their distinctive color. The massive cliff at the base of this mountain is quartzite of the Lower Cambrian Gog Group, which is also responsible for the bright red peaks and cliffs along this scenic highway. Above these quartzites are other rocks — dolostones, limestones, and shales — that were formed from sediments deposited in deep water.

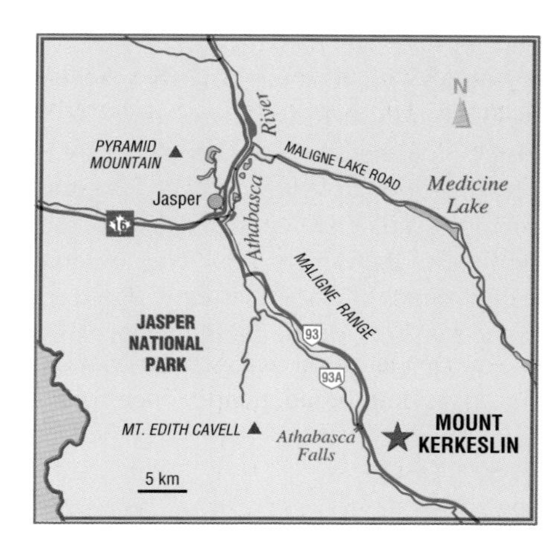

# JONAS SLIDE OF THE ENDLESS CHAIN RIDGE

David Cruden – University of Alberta

*Two slides on Jonas Ridge separated by a still stable rock mass.*

## HIGHLIGHTS

About 93 kilometres south of Jasper, the Icefields Highway was bulldozed through the Jonas Slide, a massive boulder-field of pink quartzite. Approximately 15 million cubic metres of rock broke away from Jonas Ridge, in two slides, and cleared a path nearly four kilometres long across Sunwapta Valley. Between the two rupture scars high up the slope to the northeast, there remains one million cubic metres of rock that may, one day, come tumbling down the mountain. This site is an accessible example of a major rockslide in the Rockies.

## THE STORY

Rockslides are one form of mass wasting of the earth's crust. Mass wasting is basically the downslope movement of rock debris under the influence of gravity. Slides typically occur in areas where rock layers are steeply tilting and parallel to the slope of the mountain side. Surfaces between rock layers, called bedding planes, are zones of weakness that can fail, releasing enormous slabs of rock that will rapidly slide down the mountain, occasionally on a cushion of air, from peak to valley floor.

The Rocky Mountains are a prime locality for rock slides. Throughout the Canadian Rockies there are many mountains made up of steeply dipping rock formations. Compression during mountain building thrust some rock layers up to angles between 30° and 40°, ideal angles for slide activity. Jonas Ridge, Sunwapta Peak, and Endless Chain Ridge, which can all be seen from this site, are excellent examples of mountains formed from dipping sedimentary rock layers.

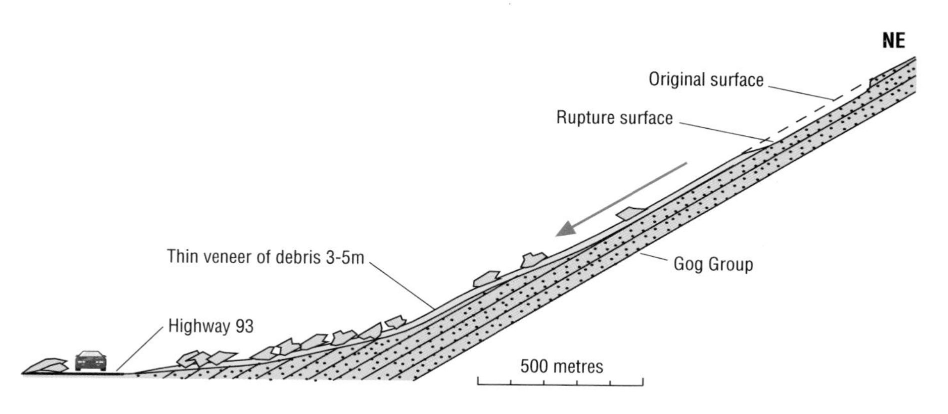


Rockslides usually occur in the spring. During the winter, water in the bedding planes expands as it freezes and slowly pries the layers apart and wedges boulders loose. Geologists believe that at Jonas Slide, downslope drainage in the spring acted as a lubricant in the rock mass that had already become unstable during the winter. Gravity took over; failure occurred along a bedding plane and the rocks tumbled down. This process is helped along by mountain slopes that have already been steepened by glacial erosion during the last Ice Age. Given the size of the trees and the amount and variety of lichens on the slide, it is estimated that it occurred several hundred years ago.

A short hike down to the Sunwapta River will bring you to the foot of the north slide, where the water breaks in rapids over pink quartzite blocks. These blocks, of the Lower Cambrian Gog Group, are used extensively as a building stone in Jasper. The Jasper Park Information Centre along Connaught Drive, for example, is built of this decorative stone.

*David Cruden – University of Alberta*

*Blocks of pink quartzite from the Jonas Slide are visible in the foreground.*

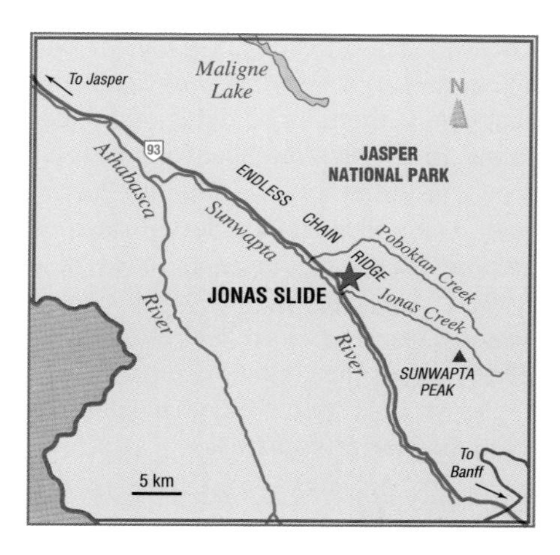

# ATHABASCA GLACIER: A Receding River of Ice

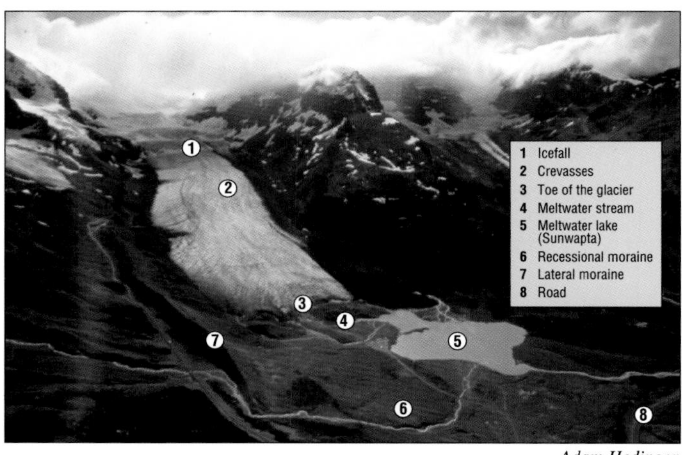


*Adam Hedinger*

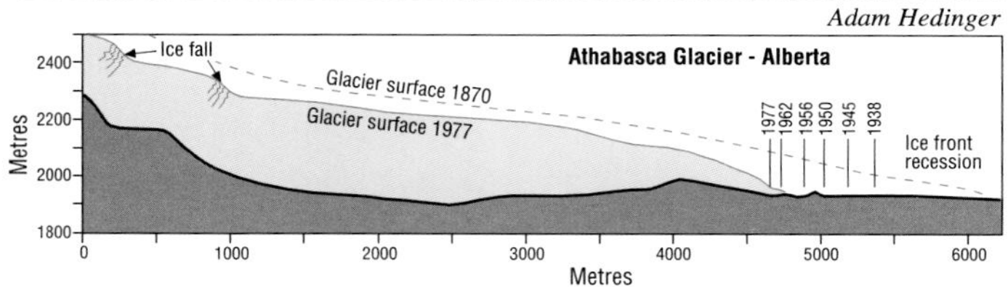


## HIGHLIGHTS

The Athabasca Glacier is the most accessible glacier in North America. A short path takes you to the glacier's toe to look at and feel some 640 million cubic metres of ice. This glacier is one of several glaciers that flow from the massive Columbia Icefields, the only point in North America where water flows to three oceans: Atlantic, Pacific and Arctic. Understanding glaciers is important to our future because they hold 75 per cent of the world's fresh water. There is an excellent interpretive centre across the highway from the glacier as well as tour buses that take you onto the glacier's surface.

## THE STORY

The Athabasca Glacier is six kilometres long and up to one kilometre wide. Because of global warming, it has melted back 1.6 kilometres since 1870 and has lost two-thirds of its volume. This means that ice at the toe of the glacier is melting more quickly than it can be replaced by snow and ice accumulating at the glacier's upper level. The road up to the glacier cuts through ridges of rock debris, where the glacier toe paused for a few years before melting again. Each ridge, called a recessional moraine, is marked with signs indicating the date it formed.

The maximum ice thickness of the Athabasca glacier is over 300 metres. It is thickest down the centre, where its downhill flow is also the quickest. This glacier flows 125 metres per year at the top, and 15 metres per year at its toe. Even at these incredible rates, it takes 150 to 200

Ron Mussieux – Provincial Museum of Alberta

*Below the toe of the glacier the exposed bedrock has been polished and striated by rock fragments that were carried in the glacier.*

years for ice to make it from top to toe! There are over 30,000 cracks in the ice, called crevasses, that develop because the ice bends as it moves.

Beside and in front of the glacier, the land is strewn with rock debris called glacial till. The glacier once carried these rock fragments beneath, within, and on top of the ice. As the ice melted, the rocks were released and deposited metres-thick over the landscape. Where the bedrock is exposed in front of the glacier, it has been polished and striated by ice-bound rock fragments that acted like sandpaper. To the left of the glacier is a 124-metre high ridge of rock debris which was formed along the edge of the ice about 100 years ago, when the glacier was much higher.

Sunwapta Lake, near the glacier's toe, is formed from melted glacial ice filling a depression scoured out of the bedrock by the once advancing glacier. It can be as much as 6 metres deep, 500 metres long, and 480 metres wide. However, with 570 tonnes of mud deposited in it each day during the summer, this lake will eventually disappear.

The rocks beneath and surrounding Athabasca Glacier are of Cambrian age, over 510 million years old. Several types of fossils, including trilobites, can be found here. Remember though, fossil collecting is not allowed in National Parks!

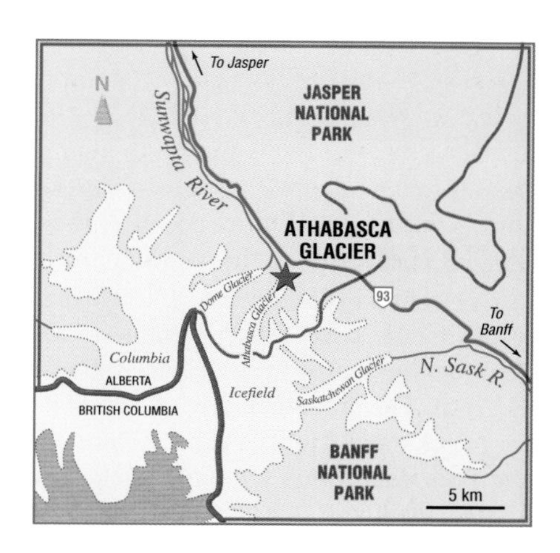

# CRESCENT FALLS: Twin Waterfalls of the Bighorn Canyon

Ron Mussieux – Provincial Museum of Alberta

*Two resistant sandstone beds form separate waterfalls.*

## HIGHLIGHTS

Crescent Falls is the descriptive name given to the two 30-metre high waterfalls at the top of spectacular Bighorn Canyon. The great diversity of sedimentary rocks and fossils in the vicinity of the falls makes this a popular field trip site for geologists. A lookout along the road to the falls provides the best view of the canyon, while the lookout at the Crescent Falls parking lot, further on, overlooks the falls themselves. A scenic trail from the parking lot winds along the canyon and passes several other small waterfalls along the way. Numerous small paths lead down from this trail into the canyon and provide access to the base of the upper falls. Beware of wet rocks!

## THE STORY

The Bighorn River begins its journey in the Rocky Mountains and flows eastwards to meet up with the North Saskatchewan River. In these mountains, the Bighorn River slowly erodes its path through dense limestone. As the river begins to flow over the softer shales and sandstones of the Foothills near Crescent Falls, erosion becomes easier and the water quickly cuts down through these sediments, forming the steep valley walls of the five-kilometre-long Bighorn Canyon.

Many Alberta waterfalls are formed where there are alternating layers of hard and soft rock. In Bighorn Canyon,

Crescent Falls and the smaller falls have been created where the river has uncovered near-horizontal and durable, resistant layers of sandstone. The water flows over the exposed sandstone but erodes the softer shales beneath. Gradually the sandstone layers become prominent ledges over which the water plummets, and falls are formed. The varying erosion rates of different rock types is called differential erosion.

The erosive power of water exposes rocks which would otherwise have remained covered. Consequently, waterfalls are popular localities for geologists to study past environments and the geological history of the province. Rocks exposed at Crescent Falls include layers of sandstone, shale, coal, and conglomerate. They provide evidence of an environment changing from floodplain (shale), to swamp (coal), and finally to a river system (sandstone and conglomerate). There are many fossils preserved in the rocks along Crescent Falls and geologists use them to estimate the age of the rocks as Cretaceous between 117 and 113 million years old.

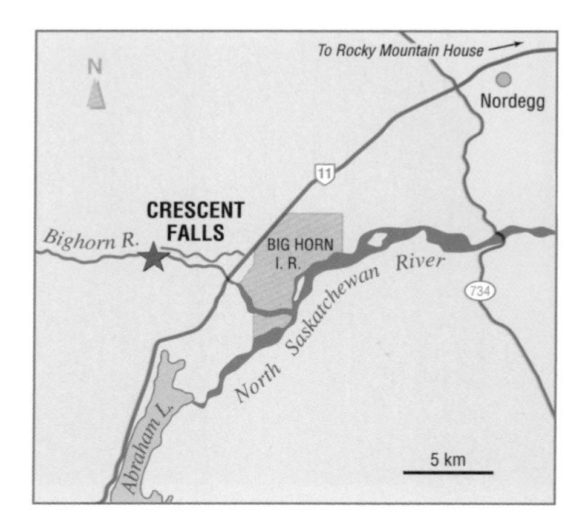

*Ron Mussieux – Provincial Museum of Alberta*

*Crescent Falls is named after the shape of the lip of the upper falls.*

# HUMMINGBIRD REEF: The Building and Fossils of a Devonian Reef

*Ron Mussieux – Provincial Museum of Alberta*

*The light grey limestone of Hummingbird Reef is covered by a fossil-rich shale.*

## HIGHLIGHTS

The Hummingbird Reef is one of the best surface exposures of a tropical Devonian reef in Alberta. Because it is exposed at the valley floor, rather than on steep cliffs in the Rockies, it is easily reached, which makes it a popular educational site for oil geologists. A rough 4-wheel drive vehicle trail leads from Ram Falls to Onion Lake. The "Headwaters Trail" leads from the east end of Onion Lake over the divide to North Hummingbird Creek. The reef is a short hike up this creek.

## THE STORY

A reef, whether fossil or modern, is made up of a community of limestone-secreting organisms that over time accumulate in mounds. Each new generation grows by cementing to the remains of the previous one and to each other, and in this manner a reef continues to grow upwards and outwards. Unlike the coral reefs of today, Devonian reefs, such as Hummingbird, consisted mainly of bulbous sponge-like creatures called stromatoporoids. Reef growth began in warm, shallow, sunlit seawater, slowly forming flat, extensive communities as small reefs grew together. After a few million years of growth outwards, either the sea level rose or the sea floor subsided, and Devonian reefs were forced to grow upwards, forming mound-like structures. Between these reefs, mud accumulated to form "shale basins," where most stromatoporoids could not survive. These off-reef muds would eventually cover the reefs themselves, effectively ending any further reef growth. Therefore, Devonian reefs

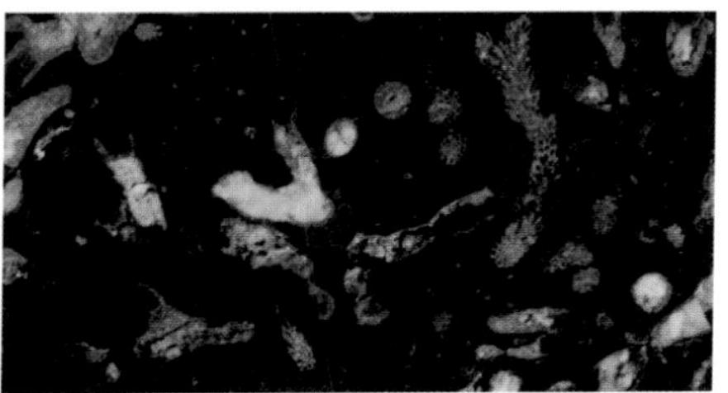

Ron Mussieux – Provincial Museum of Alberta

*Stromatoporoids (top) and branching corals (bottom) were important reef builders.*

demonstrate a sequence of fairly consistent growth stages that are characterized by specific rock types and fossils that can often be recognized from one reef to another.

There was not usually an abrupt transition from reef to off-reef, but rather a gradual interfingering. At Hummingbird Reef the actual transition has been eroded away and is occupied by North Hummingbird Creek valley. Thus, as you walk along the creek, the contrast is quite spectacular with the reef in the southeast valley wall and the off-reef in the northwest valley wall. With a valley width of only 1000 metres, the transition must have been unusually abrupt. Having both reef and basin sediments represented in such proximity has increased this reef's value as a locality to study reef rock.

Identification of the fossils of Hummingbird Reef has helped date it at about 350 million years old, making it as old as the subsurface Leduc reef. It was at least 300 metres thick and possibly tens of kilometres long! Life on this reef and in the shale basin was obviously prolific. Stromatoporoids are common in reef sediments, and beautifully preserved solitary horn corals, brachiopods, and snails can be found in the shale.

Ron Mussieux – Provincial Museum of Alberta

*Small horn corals (top) and brachiopods (bottom) lived in the muds near the reef.*

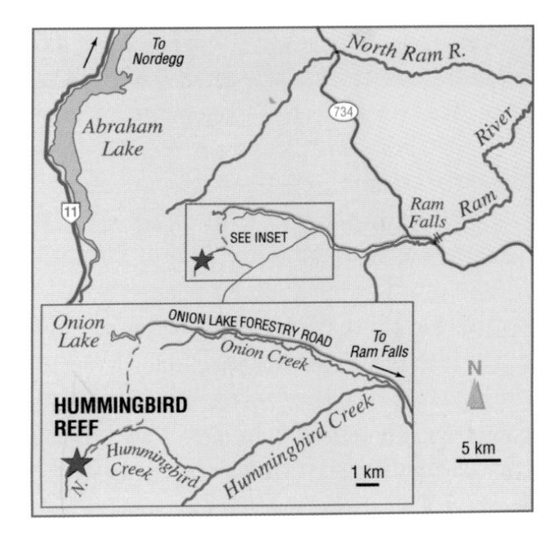

# THE FALLS AND LINEAR GORGE OF SIFFLEUR RIVER

*Ron Mussieux – Provincial Museum of Alberta*

## THE STORY

The Siffleur River has eroded through hundreds of metres of limestone and shale that were deposited between 360 and 350 million years ago, during the Mississippian Period. These rocks were originally near-horizontal sediments that were laid down on the bottom of the shallow, tropical sea that once covered much of Alberta. During the uplift of the Rocky Mountains, the layers were pushed upwards and eastward, faulted, and stacked upon each other.

The Siffleur Falls are formed where the river flows over a particularly erosion-resistant layer of limestone and quickly wears away the softer underlying shales of the Banff Formation, producing a prominent ledge over which the river plunges. Below the falls, the river continues to carve its path along this soft, weak layer — the path of least resistance. The gorge is straight because the river is flowing parallel to the hard limestone of the east wall.

Looking down the gorge from the trail, you can see that the steep east wall is the surface of a once horizontal lime-

## HIGHLIGHTS

The Siffleur River plummets over a 15-metre thick limestone ledge, through a narrow channel and finally into a straight-walled canyon below. A scenic hike, 3.9 kilometres long, that leads to the falls crosses the North Saskatchewan River by suspension bridge and passes the awesome, but dangerous, Siffleur Canyon.

stone bed, tipped up during mountain building. The bed is now tilting at a 45° angle, facing the southwest. The river has worn away the softer sediments that were once on top of this limestone layer, slowly exposing this single layer of rock. The smooth rockface is slippery when wet and has been the cause of several accidents. Looking down the river towards the northwest, you can see a continuation of the same rock structure.

The soft sediments of the Banff Formation make up the river bed at this location. Many areas where this formation is exposed in Alberta are noted as excellent fossil-collecting localities. For obvious reasons, however, we can only speculate about the variety of fossils that may be found at the bottom of this river!

*Ron Mussieux – Provincial Museum of Alberta*

*This linear canyon is eroded in a soft limestone bed.*

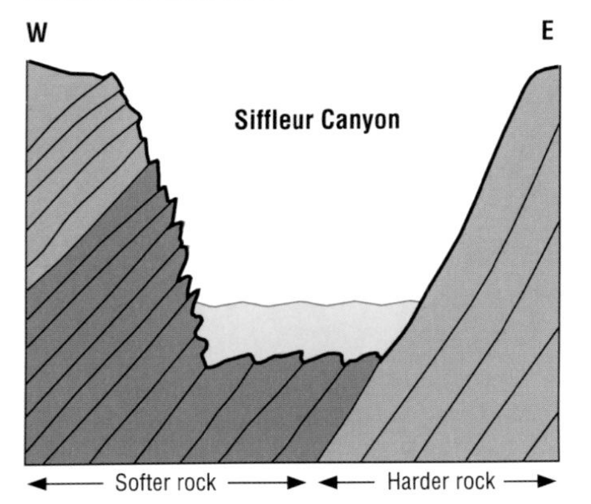


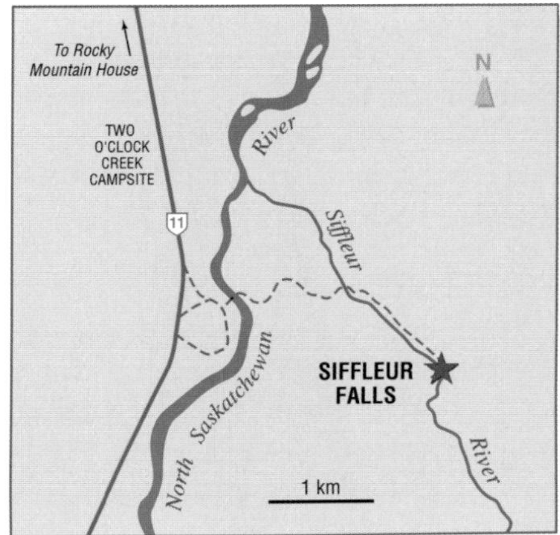

# THE BEAUTIFUL COLORS OF PEYTO LAKE

Ron Mussieux – Provincial Museum of Alberta

*View north over Peyto Lake along a broad, glacially scoured valley.*

## HIGHLIGHTS

This beautiful turquoise mountain lake is fed by water from the melting Peyto Glacier. Unlike prairie lakes, glacial lakes, such as this one, often show intense shades of green, blue, and violet that make them popular tourist destinations. The best viewpoint is at the Bow Summit Interpretive Centre along the Icefields Highway, where a short trail leads down to a spectacular view of the lake, glacier, and Mistaya Valley.

## THE STORY

Much of the charm of Peyto Lake is dependent on the sediment in the water melting from Peyto Glacier. Glacial meltwater carries vast quantities of sediments that have been ground up in and beneath the glacier. The finest of these sediments, called rock flour, resembles talcum powder and is responsible for the characteristic hues of glacial lakes, and for the milky appearance of glacial streams.

As the water enters Peyto Lake and its velocity decreases, the larger rock debris drops out, gradually building a delta out into the lake. The rock flour, however, is so tiny that it spreads out in all directions and becomes evenly distributed throughout the lake where it can remain suspended for several months. The minute flour particles reflect back only the green and blue light wavelengths from sunlight creating an opaque, vivid turquoise color.

The colors of the lake change with the seasons in response to the amount of rock flour in the water. In the winter, when the lake freezes and the glacial melting is drastically reduced, the rock

flour settles, and the water becomes indigo blue, similar to prairie lakes. With the melting of glacial ice in spring and throughout the summer, new loads of sediments are deposited and the water once again takes on the colors for which it is famous.

Peyto Glacier once extended beyond the current position of the lake. As the glacier melted and retreated, it deposited a ridge of rock debris that extends from one side of the valley to the other. This ridge, now heavily forested, holds back much of the water in Peyto Lake. Unfortunately, natural dams such as this do not last indefinitely, and eventually, the lake will either drain or be filled in by the evergrowing delta.

*Ron Mussieux – Provincial Museum of Alberta*

*Peyto Lake in spring (top) and in summer (bottom).*

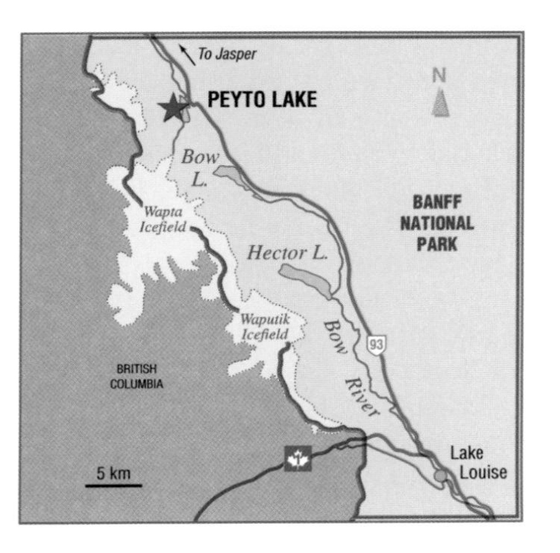

# CROWFOOT DYKE: A Rare Igneous Rock Body in Banff National Park

*Marilyn Nelson – Provincial Museum of Alberta*

*Crowfoot Dyke is a greenish-black igneous rock, a rarity in Banff National Park.*

## HIGHLIGHTS

Outcrops of igneous rocks are rare in the Canadian Rockies and therefore any that are found are carefully studied and documented by geologists. One such example is the Crowfoot dyke exposed along the Icefields Highway. A dyke is a tabular sheet of once molten rock that has flowed across the bedding of older rocks beneath the earth's surface and has hardened in a crack or fissure. The Crowfoot dyke is crossed by the Icefields Highway, one kilometre south of the Crowfoot Glacier viewpoint, or 26.8 kilometres south of the intersection with the TransCanada Highway

## THE STORY

Deep within the earth's crust are pockets of molten rock, called magma. When magma cools and solidifies, it forms igneous rocks. Magma that cools *on* the surface of the earth is called volcanic (or extrusive) igneous rock. Plutonic (or intrusive) igneous rocks are formed when magma cools *below* the surface. The texture and appearance of igneous rocks vary enormously depending on whether the magma cooled quickly on the surface or slowly below the surface.

Magma often squeezes its way into fractures in subsurface rock. A dyke often forms in near-vertical cracks, and therefore cuts across layers of older

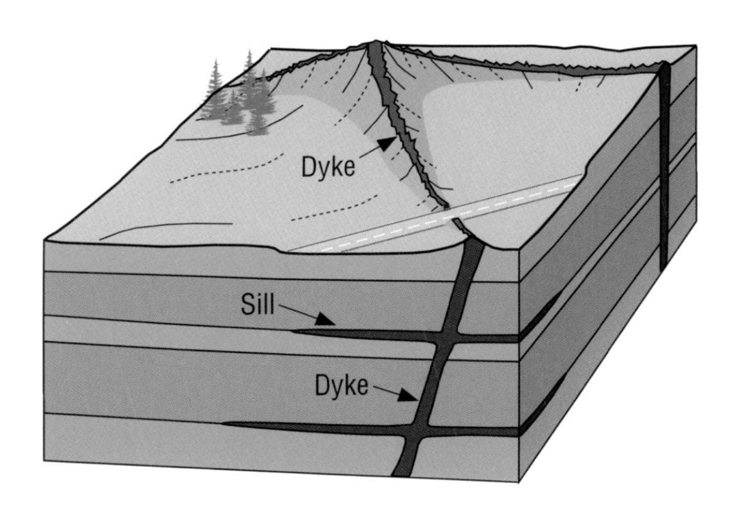


rock. It may be exposed millions of years later when younger rocks over it are eroded away, leaving the harder dyke rocks stretching across the surface of the land like a fence.

Geologists believe that the Crowfoot Dyke is 2.8 kilometres long, extending from beneath the Bow River Valley northeast as far as Helen Creek. Where it is exposed along the highway, it is 45 metres wide and 15 metres high. The dyke can be traced for 1400 metres up the hill to the northeast, but most of it is covered with vegetation and rubble. Knowing that the dyke is younger than the rock it formed in, and is older than the overlying rock, we estimate that it formed towards the end of the Precambrian, around 600 million years ago.

The size of crystals in plutonic rocks is often a clue to how quickly, and at what depth, the magma cooled. Generally, at greater depths magma cools more slowly, and crystals have more time to grow large. The Crowfoot dyke is made up of a relatively fine-grained, greenish-black rock, called diabase. The crystal size indicates the magma cooled quite quickly, probably only a few hundred metres below the surface. For about six metres on either side of the dyke, the enclosing rocks were baked, or metamorphosed, into a crumbly light grey-green rock by the magma before it cooled.

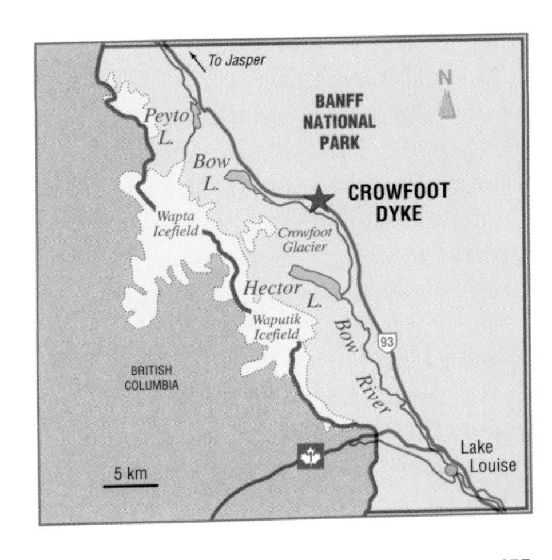

# CASTLE MOUNTAIN: Fortress of the Main Ranges

*Ron Mussieux – Provincial Museum of Alberta*

*Castle Mountain's nearly flat-lying beds are typical of the Main Ranges.*

## HIGHLIGHTS

As you travel west from Banff to Lake Louise, there is an abrupt and dramatic change in the appearance of the mountains as you go from the Front Ranges into the Main Ranges. The castle-like peaks and thick, relatively flat-lying beds of the Main Ranges, such as are seen on Castle Mountain, stand in sharp contrast to the tilted, folded, and faulted beds of the Front Ranges, seen immediately to the east in the Sawback Range.

## THE STORY

At first glance, Castle Mountain resembles a medieval fortress with its rock layers turned into cliffs, terraces, and towers by the forces of erosion. This mountain type is best developed where there are alternating beds of resistant rocks, such as limestone, and easily eroded rocks, such as shale. As the softer layers erode, the harder rocks above are undermined and break off, slowly forming a mountain of vertical walls separated by sloping ledges.

The Main Ranges are composed of older and more colorful rocks than the grey rocks of the Front Ranges. Castle Mountain consists of purple, green, and pink Precambrian and Cambrian limestones, shales, and quartzites (600 to 400 million years old) that were thrust eastward, via the Castle Mountain Fault, over the younger sediments at the base of the mountain which are now tree-covered. This fault is geologically significant as it is the feature that separates the Front Ranges from the Main Ranges at this latitude. Looking at Castle Mountain, you can see that the rocks have been folded into a broad, shallow U-shape, called a syncline. Before millions of years of erosion, this syncline was once part of the same long syncline that extended north to, and included, Mt. Kerkeslin.

Although the Main Ranges have the same southeast to northwest alignment as the Front Ranges, they lack the symmetry because of irregular drainage caused by extensive glaciation and horizontal layers. Main Range mountains are also generally higher than Front Range mountains because erosion, particularly by glaciers, wears away thick near-horizontal layers more slowly than the tilted folded layers.

On the east side of Castle Mountain is Rockbound Lake. The large depression in which the water sits was carved out by glaciers during the last Ice Age. This sort of depression, which is common in the Rockies, is called a cirque.

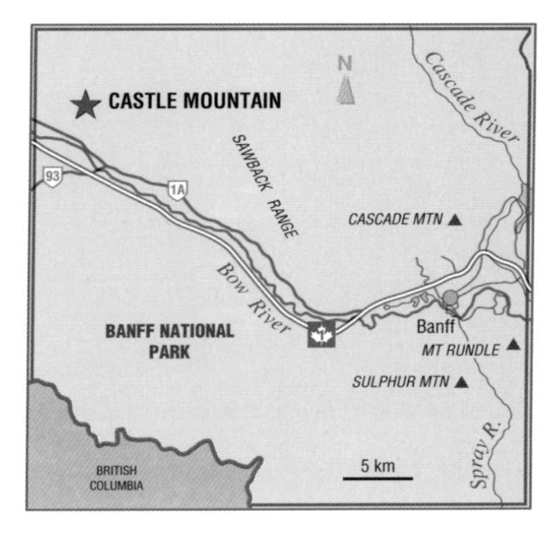

*Ron Mussieux – Provincial Museum of Alberta*

*The nearly vertical beds of the nearby Sawback Range are typical of the Front Ranges.*

# BANKHEAD: The Birth and Death of a Coal Town

*Marilyn Nelson – Provincial Museum of Alberta*

*Concrete foundations and mounds of waste coal are all that remain of the Bankhead Coal Mine.*

## HIGHLIGHTS

Bankhead, near Banff, was once the most modern and successful coal-mining town in Alberta. Its innovative method of extracting coal from Cascade Mountain became the model for many other mines. During the town's 20-year life, from 1902 to 1922, over 2.6 billion tonnes of coal were produced. Parks Canada has developed an excellent 1.1-kilometre interpretive trail through the old mining remains of Lower Bankhead.

## THE STORY

A famous Canadian geologist, George Mercer Dawson, made the first detailed survey of the Rockies in 1885 and reported outcrops of a high rank coal, called semi-anthracite, between Banff and Canmore. This discovery initiated a fierce competition for control of prime coal localities. In 1903, Canadian Pacific Railways acquired the licence to mine the coal from Cascade Mountain. The tiny town of Bankhead flourished and even became a popular tourist destination.

Cascade Mountain was a geologist's dream, with over 12 coal seams ranging from one to three metres thick in the Cretaceous-age Kootenay Formation. But it was a miner's nightmare because it was a major challenge to actually remove the coal. Each seam tilted at a 45° to 60° angle, and was folded and faulted. A seam could be traced for a distance and then would abruptly change direction or end against a fault plane. A different approach was required, and a plan was devised where the coal would be mined starting at the base of the mountain and working upwards. The plan worked but over 320 kilometres of tunnels had to be excavated into the mountain. This proved to be expensive and slow.

It soon became apparent that this unique method of extraction crushed the coal, and at least 35 per cent was wasted as coal dust. CPR solved this problem by importing pitch which was mixed with the dust to produce briquettes which would be used as fuel for steam locomotives or ships.

The First World War brought prosperous times to Bankhead as its high quality coal was needed to fuel navy warships. The coal burned well with less tell-tale black smoke than the lower ranks of coal being mined elsewhere in Canada.

By July 1922, the success story ended. A combination of post-war recession, high expenses, and union strikes closed down Bankhead forever. The buildings, houses, and equipment were moved to Calgary, Banff, and Canmore. All that remain now are crumbled ruins and mounds of abandoned coal.

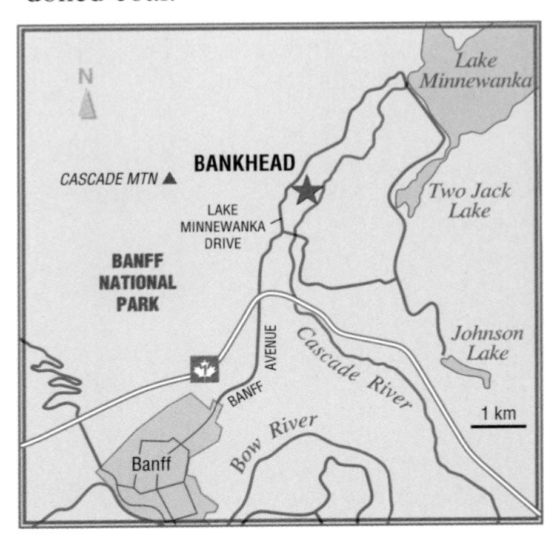

*P.840 – Provincial Archives of Alberta*

*The Bankhead Coal Mine in full operation.*

# MOUNT RUNDLE: Banff's Classic Front Range Mountain

*Grant Mossop – Geological Survey of Canada*

## HIGHLIGHTS

Majestic Mt. Rundle towers over Banff making it one of the most prominent and photographed landmarks in Banff National Park. The shape of Rundle, with its steep northeast face, tilted rock layers and gently inclined southwest-facing slope, makes it a classic example of a Front Range mountain. With an elevation of 2950 metres, it is one of the highest mountains in the Front Ranges. Excellent views of Mt. Rundle can be obtained from the Vermilion Lakes viewpoint and the Hoodoo Lookout in Banff.

## THE STORY

Front Range mountains are basically huge slabs of once-horizontal ocean floor sediments. During mountain building, the rock layers were compressed from the west, forcing them to fold and buckle. Eventually, the layers fractured into slabs, called thrust sheets, which slid eastwards along thrust faults to their present locations. Each thrust sheet was tilted up and successively stacked on top of the sheet in front, creating a symmetry similar to shingles on a roof. Viewed from the air, the Front Ranges are a series of near-parallel thrust sheets, or mountain ranges, all with a northwest to southeast alignment. Each thrust sheet, as seen in Mt. Rundle, has one gentle slope, one steep slope, and a thrust fault at its base. Mt. Rundle is part of the Rundle Thrust Sheet which also includes Cascade Mountain and the Three Sisters.

Mt. Rundle is composed of the three rock formations that are familiar to geologists because, through thrust fault-

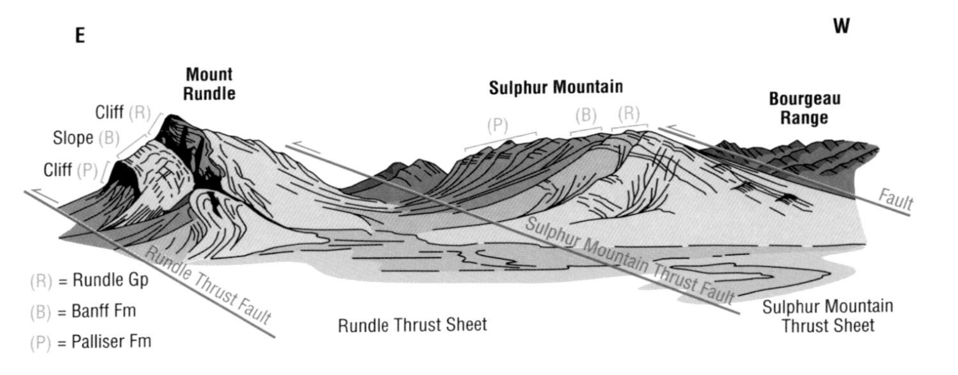


ing, they also form many other Front Range mountains. These rocks consist of an alternating sequence of limestone-shale-limestone which forms a distinctive cliff-slope-cliff pattern. This pattern can be seen perfectly from Hoodoo Lookout. The peak is formed of thick cliff-forming Rundle limestone that was deposited during the Mississippian Period. Below this are limestone and shales of the Mississippian Banff Formation. This formation, which forms ledges and slopes, can locally reach 1200 metres in thickness. At the base of the mountain is a massive limestone cliff of the Devonian Palliser Formation. Steep, thick cliffs of this formation are very common in the Rockies and were deposited in an environment similar to the enormous limestone banks of the modern Bahama Islands. This same sequence of rock layers is repeated numerous times from east to west and can be seen across the valley in Sulphur Mountain.

Erosion over millions of years has carved this complex pattern of rocks into mountains where the layers are erosion-resistant, and into valleys along beds of softer rock or along fault lines. The southwest slopes of Front Range mountains are particularly vulnerable to erosion as entire layers can break free and slide to the valley floor. Thus, massive rock slides are common in this type of mountain. Someday perhaps, Mt. Rundle will also suffer this fate.

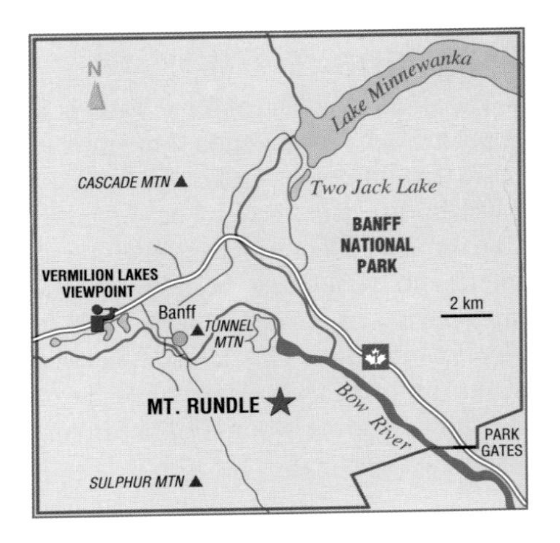

# BOW FALLS: Phosphates in the Cliffs

*Marilyn Nelson – Provincial Museum of Alberta*

## HIGHLIGHTS

The broad cataract of Bow Falls is a popular tourist attraction in Banff, and is best viewed from the lookout near Spray River bridge. It is the cliff on the north side of the falls, however, that is of interest to geologists, because it contains layers of phosphate-rich rock, a rare commodity in Canada. Several steep, dirt paths lead down to this side of the river opposite to the Banff Springs Hotel.

## THE STORY

At Bow Falls, the river flows along the boundary between two rock formations, which offers a path of least resistance. Upstream, the cliff on the left side is composed of sandstone and siltstone of the Spray River Group, which was deposited during the Triassic Period, about 225 million years ago. These rocks are soft, easily eroded by the flowing water, and extend into the river bed as jagged steps, producing the rapids and falls. The siltstone is an excellent building stone, and can be seen on the front of the Banff Springs Hotel, on the Banff Administration Building, and even on buildings in Calgary. Today, there are two operating quarries mining the Spray River siltstones, both located near Canmore.

The older rocks that make up the steep, smooth cliff on the right side are quartzite, shale, and layers of phosphate rock of the Ishbel Group, named after Mt. Ishbel (part of the Sawback Range). These rocks were deposited during the Permian Period, about 250 million years ago. The phosphate can be recognized as a thin, dense, black, and quite hard layer of rock. Occasionally tiny purple fluorite crystals may be found nearby. Even if it were not located in a national park, this phosphate deposit is too thin to be mined economically.

Why are phosphates important? They are used in the production of fertilizers and are therefore of particular importance to the agriculture industry in Alberta. All the ingredients of fertilizers are found in Canada except the phosphates which must be imported at great expense, mainly from the United States. Thus, the discovery of these phosphate deposits in the Canadian Rockies, although not of commercial quality, is promising. The individual who does discover extractable phosphates will be fortunate indeed!

Phosphates are forming today as nodules on the seafloor. Some of the phosphates are believed to be derived from phosphorous-rich fish teeth, bones, and scales, as well as the shells of tiny ostracods (a type of crustacean).

Look for fossil ripple marks on the cliffs on the left side of Glen Avenue as you drive down to the falls. These were formed in the shallow water of an ancient ocean, and preserved as the sand and silt was hardened to rock. They were later raised to their present position during the formation of the Rocky Mountains.

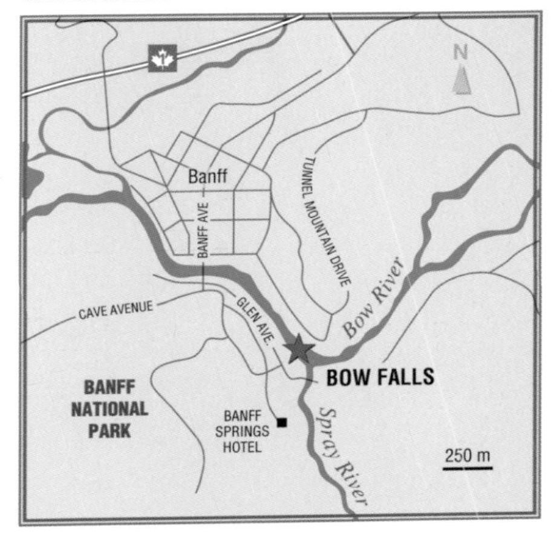

*Marilyn Nelson – Provincial Museum of Alberta*

*Steeply dipping rock layers on the east side of the falls contain a thin black layer of phosphate.*

# GRASSI LAKES: A Tropical Reef in the Mountains

*Ron Mussieux – Provincial Museum of Alberta*

*View down the near-vertical limestone face to Grassi Lakes.*

## HIGHLIGHTS

The limestone cliff above Grassi Lakes is an ancient reef that grew in a tropical ocean between 380 and 360 million years ago during the Devonian Period. During mountain building millions of years later, this reef was raised thousands of metres and was exposed in the mountains. Devonian reefs beneath the Plains store almost two-thirds of Alberta's conventional oil and gas reserves. Studying similar reefs exposed on the surface contributes towards the understanding and development of future reef-based petroleum reservoirs.

## THE STORY

A Devonian reef is a mound-like rock structure built by sedentary organisms, such as corals or stromatoporoids, and is often enclosed in a different rock type. During the Middle and Late Devonian Period much of Alberta was covered with a warm, shallow ocean that teemed with life. The ocean floor was dotted with reefs, many of which would become major oil producers millions of years later. Like the Grassi Lakes reef, these reefs consisted of bulbous creatures, similar to sponges, called stromatoporoids. Stromatoporoids were the most abundant reef builders during Devonian times, making up over 90 per cent of reef life. They grew in stationary colonies as crusts or layers, and each new generation grew on top of or beside previous layers. Stromatoporoids covered hundreds of square kilometres of ocean floor, and formed reefs nearly 300 metres high! Towards the end of the Devonian Period, however, most died

Adam Hedinger

*View of Grassi Lakes reef and trail from the lake level. Notice the cavernous nature of the reef limestone.*

*Fossil stromatoporoids are weathered out leaving large holes in the limestone.*

Adam Hedinger

off and Alberta's reef building age was over.

The stromatoporoid reefs were eventually buried by other sediments, and over millions of years were turned into rock. Deep beneath Alberta's Plains there are many reefs formed of stromatoporoid fossils. These blob-shaped organisms are important to Alberta's oil industry and have brought to the province much of its petroleum wealth. Why? Most of their structure was destroyed in the years following their burial, resulting in the hollow irregular cavities we find today. These cavities provided the ideal place for petroleum to accumulate, making stromatoporoid reefs superb oil and gas reservoirs. The Grassi Lakes reef has abundant large cavities, many reaching several centimetres in diameter. Because other Devonian subsurface reefs, such as Leduc and Nisku, are similar to this one, it is easy to understand why geologists call stromatoporoid reefs "Alberta's billion dollar rock!"

Although stromatoporoids are a significant part of Grassi Lakes reef, they are usually difficult to recognize because of their typically poor preservation. There are also other loose fossils, such as corals, brachiopods, and snails, that can be found in the rubble above the upper lake.

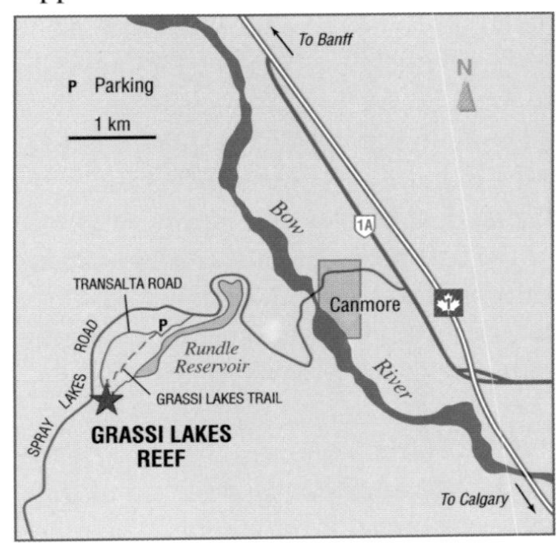

# MOUNT YAMNUSKA: First Grey Wall of the Front Ranges

Samuel Nelson – University of Calgary

*The trace of the McConnell Thrust Fault can be seen in the break of the slope where snow has accumulated. Massive older Cambrian limestones lie on top of younger, soft Cretaceous sandstones and shales.*

## HIGHLIGHTS

Mount Yamnuska, the first major mountain of the Front Ranges west of Calgary, abruptly rises 1000 metres above the rolling landscape of the Foothills. Along the base of its vertical cliffs runs the McConnell Thrust Fault, a geologically significant fault because it is the dividing line between the Foothills to the east and Front Ranges to the west. This fault, recognized in 1887 by R.G. McConnell, a geologist for the Geological Survey of Canada, was one of the first major thrust faults to be identified in the Rocky Mountains and can be traced for 400 kilometres along the east side of the Front Ranges.

## THE STORY

Mount Yamnuska is made up of only two rock units, or formations. The rocks forming the 350-metre high cliff are hard, resistant limestones of the Eldon Formation that were deposited during the Cambrian Period, between 525 and 515 million years ago. Beneath these are much younger mudstones and sandstones of the Belly River Formation, that were deposited on land around 75 million years ago, during the Cretaceous Period. These younger sediments, and those of the surrounding foothills, are much weaker than the overlying tough limestones. Thus, as the Foothills erode away, the imposing cliff of Mount Yamnuska endures, leaving behind the vertical mountain front so popular with rock climbers.

The most notable geological feature of Mount Yamnuska is that older rocks sit on top of much younger rocks. This scenario, of older over younger rocks, is

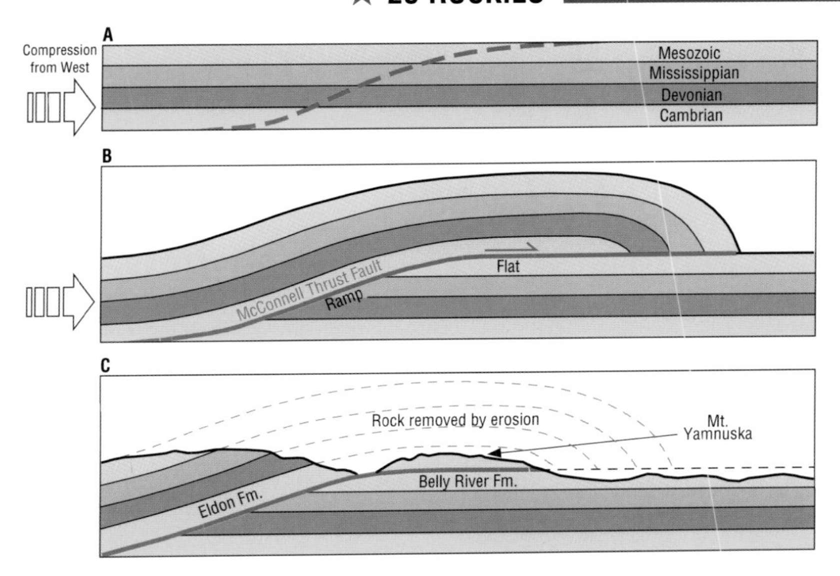


characteristic of the Rocky Mountains. During the building of the Rocky Mountains, compression from crustal plate collisions to the west forced thick layers of ocean floor sediments to buckle and break into immense slabs of sedimentary rock, called thrust sheets. These sheets were thrust up and pushed northeastwards along enormous thrust faults, such as the McConnell Thrust, and over younger rock formations. Reconstruction of the rock formation (before folding and faulting) indicates the Eldon Formation was displaced, or moved, along the McConnell Thrust 45 kilometres east to its present location at Mt. Yamnuska. This site is particularly interesting because, even though this massive limestone block slid along the fault over the much weaker sandstones and shales of the Belly River Formation, it did not disturb the underlying rocks. Some geologists have suggested that water in the pores of the sandstone created enough pressure that the rock was not deformed by the over-riding limestone. Millions of years of erosion by water, ice, wind, and gravity have sculpted the mountains we see today.

Thrust faults in the Rocky Mountains have a characteristic outline. The initial faulting occurred along a horizontal "flat." The layers of rock then slid up a "ramp" and finally flattened out into another "flat." Looking at Mt. Yamnuska from a distance, the ramp and the flat are clearly visible.

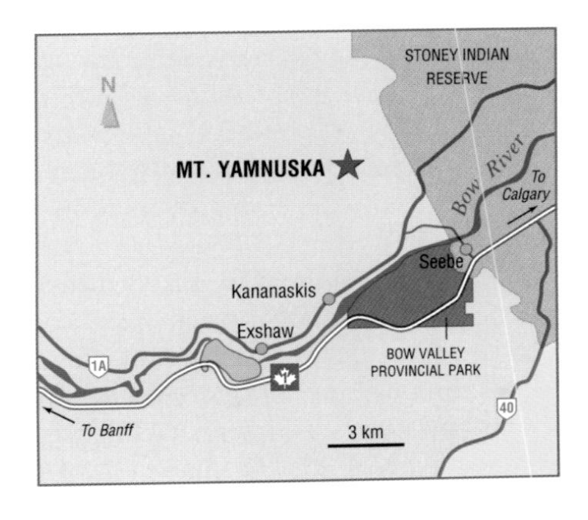

# DRUMLINS: The Streamlined Hills of Morley Flats

Marilyn Nelson – Provincial Museum of Alberta

*A Morley Flats drumlin.*

## HIGHLIGHTS

At Morley Flats, 42 kilometres west of Calgary, there is a group of remarkable hills that glacial geologists call drumlins. Drumlins are elongated oval hills that form beneath flowing glaciers. They are reliable indicators of the direction of flow of long-melted glaciers and therefore help us to analyse the glacial history of Canada. The drumlins at this location are well preserved and easily viewed from the TransCanada Highway.

## THE STORY

The drumlins in Morley Flats consist of rock fragments of varying sizes. As glaciers flow, they pluck up rock fragments from bedrock and carry the debris to be deposited elsewhere. Geologists call this type of deposit glacial till. As the rock fragments begin to accumulate at the base of a glacier, the ice moulds them into streamlined, smooth hills called drumlins. As we can not actually observe a drumlin being formed, there are many theories on how it happens. Most theories agree, however, that meltwater beneath a glacier plays an important role.

Along Morley Flats, there is a cluster of drumlins, called a "drumlin field." These drumlins, which run parallel to the TransCanada Highway, formed during the last Ice Age. They are irregular, elongated ridges that rise above the surrounding plains, many up to 25 metres high. They are from 75 to 300 metres wide and from 300 to 400 metres long. Viewed from the air, they give a distinct impression of ice motion. Most are

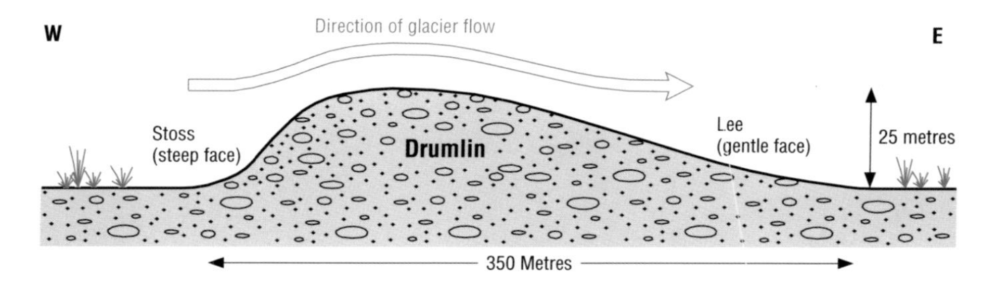


wooded over, but there are two classic drumlins that can be seen from the highway.

Drumlins tell a story about the direction of ice flow; the blunt, steep end points in the direction from which the ice flowed (upstream) and the gentle, tapered end points downstream. They are always elongated parallel to ice movement and therefore offered a minimum of resistance to the glacier riding over them. Looking at the Morley Flats drumlins, we can see that the west ends are much steeper than the east ends. Therefore, we know the Bow Valley glacier was flowing from the mountains in the west and was heading eastward.

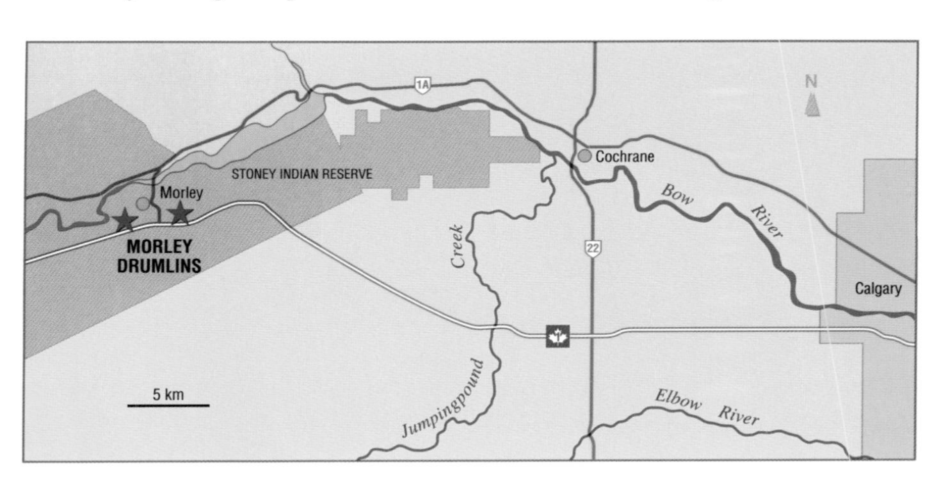

# THE STONE PATTERNED GROUND ON PLATEAU MOUNTAIN

Ron Mussieux – Provincial Museum of Alberta

*Patterned ground on the surface of Plateau Mountain.*

## HIGHLIGHTS

Plateau Mountain is capped with a broad, treeless, 14-kilometre-long plateau which gives the site its name. The surface of the plateau is "decorated" with patterned ground, where frost and gravity have rearranged rock fragments to produce distinctive geometric shapes. This is one of the best areas in Alberta to view this phenomenon and can be accessed by a steep hike up the gas well road. This site has been designated a Natural Area.

## THE STORY

Patterned ground is a general term for a surface where rocks and finer sediments have been arranged into somewhat symmetrical, geometrical patterns. On Plateau Mountain, larger rocks have accumulated to form polygons and circles with finer sediments in the centres. Patterned ground is best developed in regions of the world where there is an intense freeze-thaw cycle. When the ground surface freezes, it contracts leaving deep cracks or wedges, occasionally with a symmetrical pattern, somewhat like mud cracks. This, then, is the beginning of patterned ground. A continuing freeze-thaw cycle heaves the earth into mounds and separates larger rock fragments from the smaller ones. Larger rocks work their way up to the surface more quickly. Once on the surface, more frost-heaving below the surface in combination with gravity may push and roll the rocks into the enlarged cracks. Over time, enough rocks become wedged in the cracks to produce the patterned effect.

The polygons on Plateau Mountain are up to four metres across and most are developed on the top of the plateau, where the ground is almost horizontal.

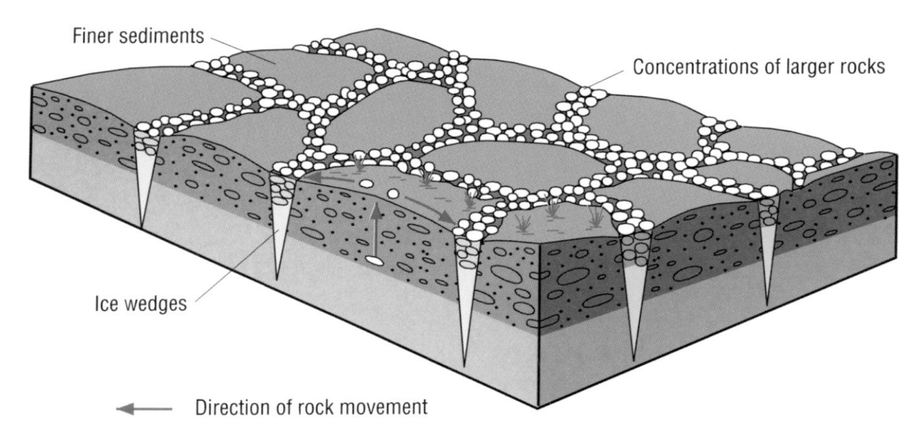


As the ground slope increases towards the sides of the plateau, the polygons become distorted by gravity and slowly become circles and eventually stripes. Stone stripes at the steep sides of the plateau will channel water running off the top which enables the sides to remain stable and steep, without slumping.

In some areas of the world, where fossil patterned ground has been discovered, it helps to reconstruct past climates and environments. The patterned ground of Plateau Mountain was likely formed during the last Ice Age when many of the surrounding valleys were filled with ice. The plateau remained unglaciated, however, and in the cold, harsh climate, frost action slowly created the surface we see today. Looking at the mountain side, you can see large depressions hundreds of metres across. These were formed by glacial erosion, and are called cirques.

The elevation of the plateau is 2500 metres and has a climate resembling the Arctic. Interestingly, the vegetation pattern is closely related to the patterned ground as the rock borders of the polygons have a different microenvironment than the finer sediments of the interior.

*Ron Mussieux – Provincial Museum of Alberta*

*Stone patterned ground showing finer pebbles in the middle and large lichen covered rocks towards the edges.*

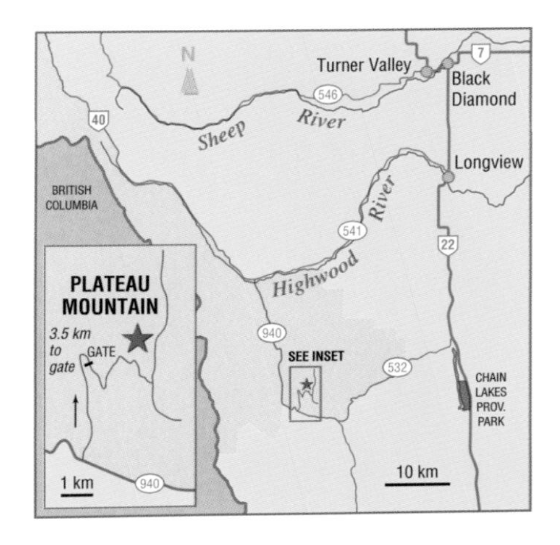

# CHAPTER 6: CALGARY AND AREA

*Cascading stream at Big Hill Springs Provincial Park.*

*Ron Mussieux – Provincial Museum of Alberta*

# CALGARY AND AREA

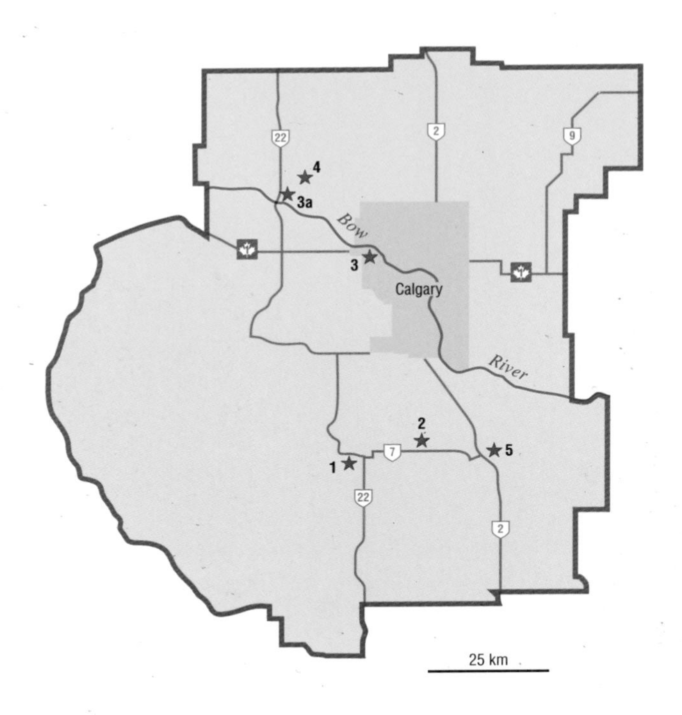


- ★ **1** Turner Valley Oil and Gas Field
- ★ **2** Okotoks
- ★ **3** Calgary Building Stone Quarries
- ★ **4** Big Hill Springs Provincial Park
- ★ **5** Gypsum Localities at Aldersyde and Chain Lakes

# TURNER VALLEY GAS FIELD AND HELL'S HALF ACRE

*Ron Mussieux – Provincial Museum of Alberta*

*The Royalite gas plant at Turner Valley is a Provincial Historic Site. Note the burning natural gas — a small reminder of gas flaring at Turner Valley's "Hell's Half Acre."*

## HIGHLIGHTS

The 1914 discovery of the Turner Valley oil and gas field, 45 kilometres southwest of Calgary, was Alberta's first major oil strike. In many ways, Turner Valley provided the impetus for the development of Western Canada's multi-billion dollar petroleum industry. The historical importance of this oil field was recognized in 1989 when the Royalite gas plant, built in the early 1920s, was designated a Provincial Historic Site by the Government of Alberta.

## THE STORY

In the early 1900s, ranchers began to notice natural gas seeps along the Sheep River. A local rancher, William Herron, finally urged Calgary businessmen to drill for oil on his property. In 1914, the Calgary Petroleum Products Company Dingman No.1 well blew in with natural gas and a light, clear liquid petroleum, called naphtha. This discovery triggered an oil boom and later that year 226 oil companies were listed in Calgary's city directory! The boom was short-lived, however, as World War I brought development to a grinding halt.

In 1921, the Royalite Company took over drilling and in 1924 they struck a major oil and gas reservoir in a thick layer of limestone — this started the second boom in the valley. Most of the gas recovered was not used, however, but flared off in a nearby ravine known as Hell's Half Acre because of the constant fire. At night, the light from these flares could be seen as far away as Calgary.

The last and most prolific economic

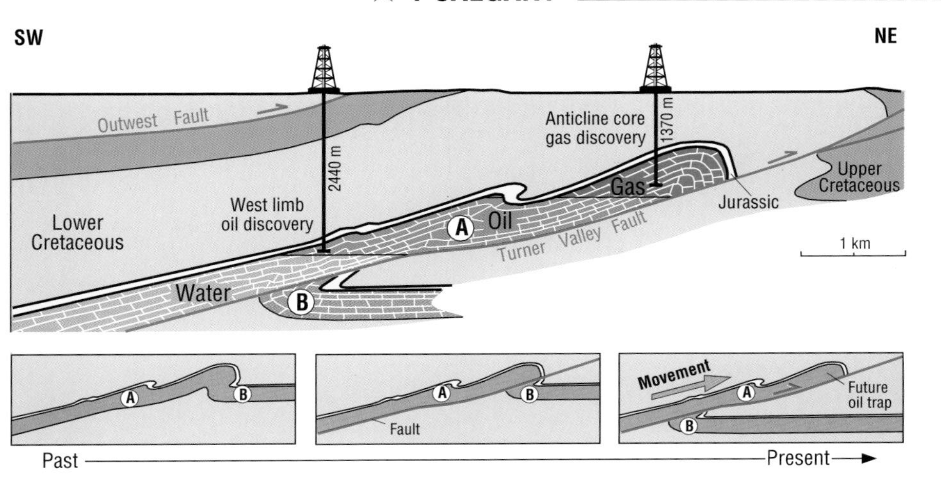


boom began in 1936 when Turner Valley Royalties hit crude oil in the southwest limb of the anticline. The Turner Valley field hit peak output in 1942 with 10 million barrels, and production declined steadily after that.

The oil and gas of the Turner Valley field are concentrated in a structure called an anticline trap. An anticline occurs when layers of sediments are folded into an arch shape. This type of trap is the simplest and easiest for geologists to recognize and tends to occur mostly in the Foothills. At Turner Valley, the petroleum has simply accumulated in the crest of the anticline consisting of porous Mississippian-age limestone. Because there is an overlying impermeable caprock, it was trapped and prevented from migrating. Looking at the diagram, you can see the anticline was faulted and moved eastwards to its present location.

Turner Valley was exploited in such a manner that an estimated 88 per cent of the oil remains in the reservoir. Crowding of wells and the flaring of natural gas depleted the gas pressure in the reservoir and the remaining reserves are now unrecoverable by present-day technology. The lessons learned at Turner Valley led to the formation of the Energy Resources Conservation Board, a government agency charged with the responsible recovery of our petroleum resources.

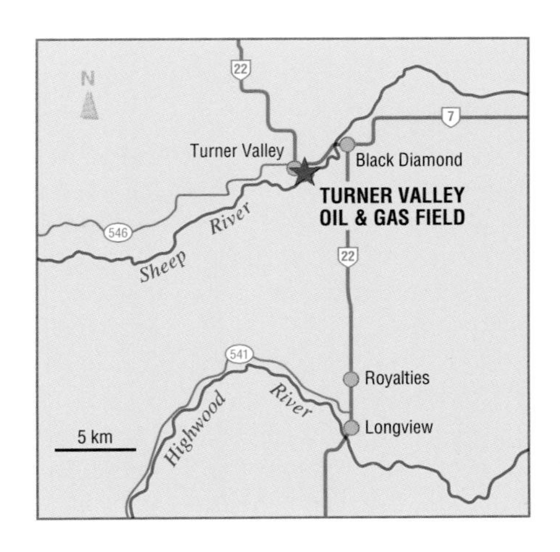

# THE OKOTOKS "BIG ROCK": A Glacial Hitchiker

*Ron Mussieux – Provincial Museum of Alberta*

## HIGHLIGHTS

This house-sized quartzite block is the largest and best known "glacial erratic" in Alberta. It is 41 metres long, 18 metres wide, 9 metres high, and weighs 16,500 tonnes! A glacial erratic is a rock fragment that has been moved by glacial ice and deposited many kilometres away from its original position. This erratic has been designated as a Provincial Historic Site.

## THE STORY

Geologists know that Big Rock is not of local origin because it is composed of a hard rock, called quartzite, rather than the local soft sandstones. The silts, sand, and pebbles that make up the quartzite were deposited on a shallow sea floor between 570 and 540 million years ago. Over millions of years, they were buried under hundreds of metres of sediments. The weight and heat generated by the overlying sediments slowly cemented and compacted the sands, silts and pebbles into hard, durable quartzite.

Between 150 and 50 million years ago, during the formation of the Rocky Mountains, the quartzite beds were pushed to the east and thrust upwards thousands of metres. Quartzite is a common rock making up much of the Main Ranges in the Rocky Mountains.

Geologists believe that approximately 18,000 years ago, rockslides in the mountains of the Jasper area dropped large blocks of quartzite onto the surface of the passing Athabasca valley glacier. The flowing glacier slowly carried its debris eastward to the plains. It is amazing a block of this size could be carried this distance on the surface of a glacier, be lowered to the ground as the ice melted, and still remain intact.

Big Rock is part of a 580-kilometre-long chain of erratics. This chain, called

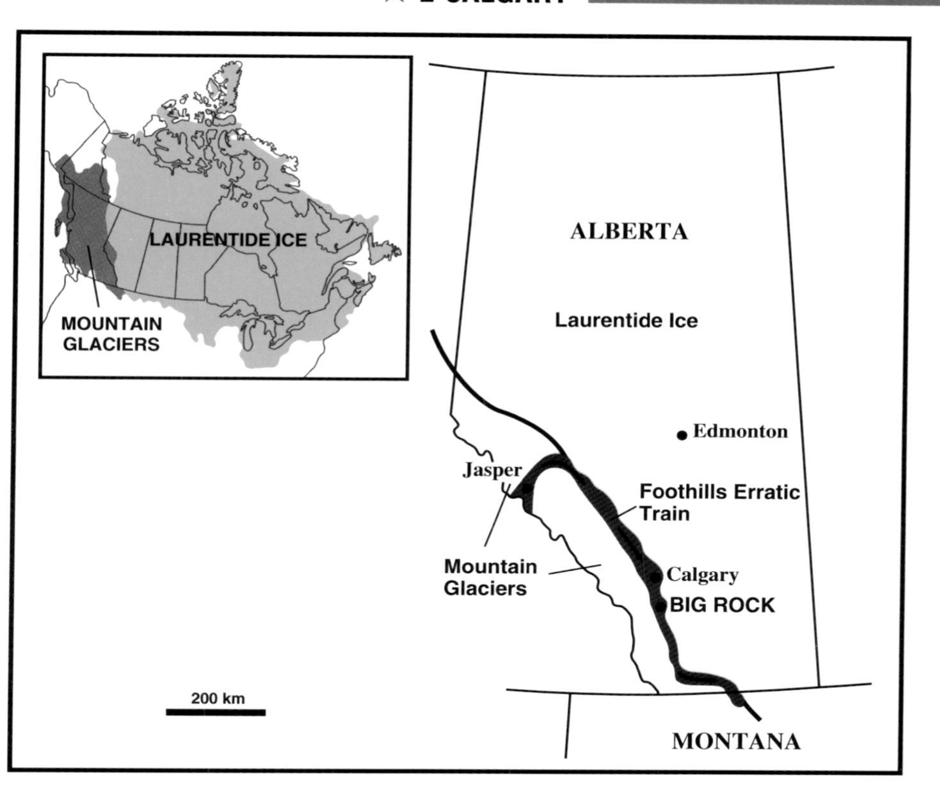


the Foothills Erratics Train, extends from beyond the McLeod River in the north to the Canada-United States border in the south. The Foothills Erratics Train is narrow, ranging in width from 22 kilometres to less than one kilometre. Likely, as the erratic-bearing glacier flowed out of the foothills, it collided with a continental glacier. The mountain glacier was deflected southwards and flowed beside the continental glacier. Other valley glaciers emerging from the mountains also joined this stream of ice. The erratics were confined to the now elongated Athabasca glacier and this would account for the length and narrowness of the Foothills Erratic Train.

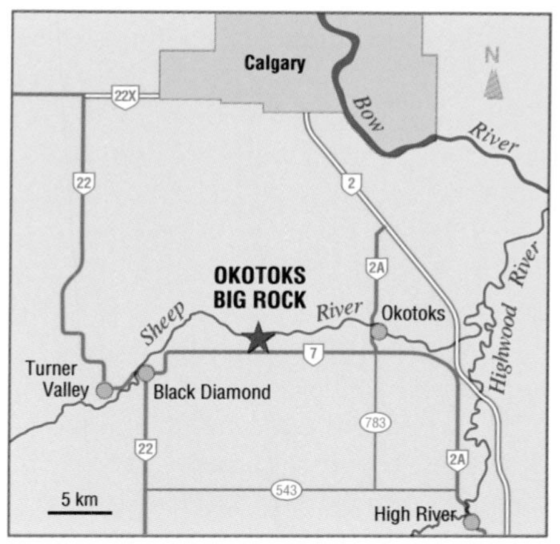

## CALGARY: The Sandstone City of the West

*Ron Mussieux – Provincial Museum of Alberta*

*Thick Paskapoo sandstone beds exposed in the Edworthy Quarry, Edworthy Park, Calgary.*

## HIGHLIGHTS

In November of 1886, many wooden buildings along Calgary's main street were destroyed in a devastating fire and Calgarians decided to rebuild with a safer, and more durable, building material. Townspeople noticed the potential of the massive sandstone outcrops along the Bow and Elbow river valleys and they opened the first quarry in 1886. At the height of the "Sandstone Era," fifteen quarries operated in and around Calgary and the beautiful golden-brown sandstone buildings gave the city its deserved title — The Sandstone City of the West. Traces of most of the quarries in the Calgary area were destroyed by city expansion but the remains of the Edworthy Quarry can be seen immediately west of the entrance road to Edworthy Park. The debris from several other small quarries can be seen on private land from the Big Hill Springs Road just north of Cochrane.

## THE STORY

The sandstone outcrops belong to the Paskapoo Formation of Paleocene age, 65 to 58 million years old. They were deposited as river sands that had been eroded from the Rocky Mountains to the west. These sands were eventually buried under hundreds of metres of younger sediment, cemented with clay and carbonate minerals, and then slowly exposed by erosion. Today, the Paskapoo Formation is found from Pincher Creek in the south to east of Grande Cache in the north, and extends from the east side of the Foothills to the Hand Hills, east of Drumheller.

Following the opening of the first

quarry in Calgary, the demand for the sandstone mushroomed, and several quarries opened from Fort Macleod in the south, to Entwistle on the Pembina River (see page 90) in the north. Near Calgary, the towns of Sandstone, Brickburn, and Glenbow were established near quarries to house the workers. Skilled stonemasons were brought from Europe and, by 1890, over 50 per cent of Calgary's tradesmen were working in the sandstone industry.

The sandstone was used to build large government buildings, schools, churches, and private buildings; as demand increased, it was shipped to other cities in Alberta. Perhaps the most significant sandstone building is the Legislature Building in Edmonton which was built mainly of rock quarried at Glenbow, just west of Calgary. The Sandstone Era came to a close in 1915 with the outbreak of the First World War, although there was a short-lived revival of quarrying north of Cochrane in the early 1930s.

Paskapoo sandstone was a soft sandstone, easy to carve, and came in a range of brown colors, with overtones of yellow, buff, grey, and even blue. Unfortunately it also had a number of disadvantages. Its hardness was highly variable and only the weathered surfaces close to the ground surface were soft enough to be worked. The rock also contained clay inclusions, had irregular bedding planes, and was susceptible to damage from wetting and freezing. Ultimately, it could not compete with the more durable Tyndall limestone of Manitoba.

NA 3267-53 Glenbow Archives

*Stonecutters and a 10-tonne sandstone block in a quarry near Calgary.*

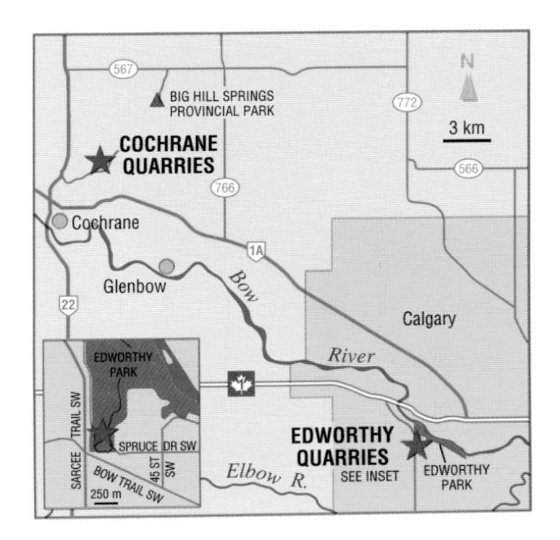

# THE BEAUTIFUL TUFA DEPOSITS OF BIG HILL SPRINGS PROVINCIAL PARK

*Ron Mussieux – Provincial Museum of Alberta*

*A mound of tufa, a fresh water limestone, marks the site of a former spring.*

## HIGHLIGHTS

As early as 1935, J.A. Allen, Alberta's first resident geologist, recommended that the area around Big Hill Springs be set aside for protection as a park. He wrote: "This short valley containing tufa deposits is one of the most beautiful spots in Alberta east of the Rocky Mountains, and I recommend that it be preserved as a park reserve for the pleasure of visitors." Today, this picturesque 31-hectare park, located 16 kilometres northeast of Cochrane, is a secluded day-use area where there is an interpretive trail, scenic viewpoint, and historic fish hatchery. The centre of the park is a steep-walled valley where Bighill Creek trickles through and cascades over tufa ledges to form numerous small waterfalls.

## THE STORY

The valley of Bighill Creek began to form over 10,000 years ago when the continental ice sheets covering Alberta began to melt and retreat, water beneath and around the edges of the ice gouged out 80 metres of the underlying bedrock, gradually carving this little valley.

The most interesting geological features in this park are the mounds of tufa seen near the bottom of the valley. Tufa is a spongy, porous limestone (calcium carbonate) that is soft when wet and hard when dry. It is deposited by lime-rich spring water that seeps from the sandstone escarpment on the west side of the valley and down into the grassy ravine below. The natural carbonate cement in the sandstone is the source of the lime found in the spring water. As the water comes into contact with the air, it warms and the calcium carbonate is precipitated out, often incorporating leaves and plant stems. Slowly, the tufa mounds are built up. The deposits extend along the valley floor for about 260 metres and in some places have built waterfalls. The waterfalls can be quite colorful because of the vividly colored algae that often grow on the creamy white tufa.

Over the years, blocks of this tufa have been quarried and used for rock gardens and decorative walls, and in buildings in Edmonton, Banff, and Calgary. In 1935, the Flesher Marble and Tile Company, Calgary, even used some of the stone for terrazzo work. Fortunately for the preservation of this park, the tufa was too soft and porous to justify an ongoing quarrying industry.

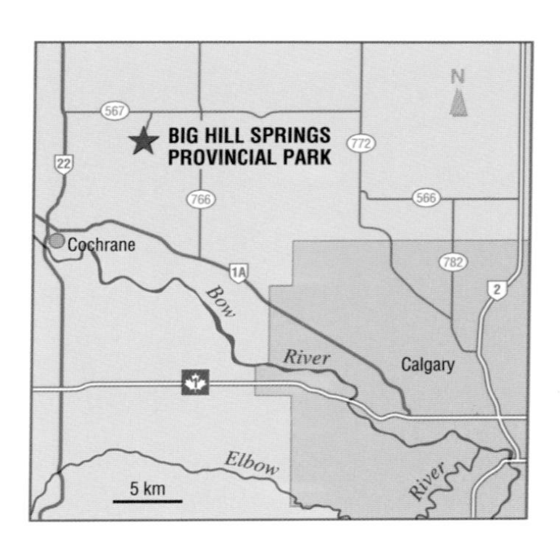

Ron Mussieux – Provincial Museum of Alberta

*This tufa mound makes an excellent shelter as well as good building stone as seen in this fireplace.*

# ALDERSYDE AND CHAIN LAKES GYPSUM CRYSTALS

Ron Mussieux – Provincial Museum of Alberta

*A transparent single crystal of gypsum.*

## HIGHLIGHTS

Both the Chain Lakes and Aldersyde areas produce beautiful gypsum crystals that can be found by collectors. These areas are documented sites of quality gypsum crystals in Western Canada, and specimens collected from here are on display in the mineral gallery at the Provincial Museum of Alberta.

## THE STORY

The best locality to look for gypsum crystals in the Chain Lakes area is on the west side of Highway 22, about 7.5 kilometres south of the intersection with secondary road 533. This area is in Happy Valley, near the junctions of Langford Creek and South Willow Creek. Here, gypsum crystals occur as perfect, flattened prisms that are transparent and colorless with an average length of three centimetres. The most distinctive property of gypsum is its softness: a fingernail can scratch its surface. This softness, however, does pose collecting and handling problems. Careful consideration has to be given to wrapping, handling, and transporting these delicate crystals. It is best to search for this mineral below ground level (as deep as 150 centimetres) as it tends to become damaged once it is exposed to the elements at the surface.

Gypsum, chemically written as $CaSO_4 \cdot H_2O$, is a calcium sulphate that contains some water within its crystal structure. At Chain Lakes, the crystals grow in a predominantly clay soil, with some sand and silt. These sediments were eroded from the Rocky Mountains and then transported via flowing glaciers and are therefore high in sulphur (from the shale) and calcium (from the limestone). Clay acts as a natural barrier to the movement of groundwater. The water slowly circulates through the clay-rich layers where high levels of calcium and sulphur steadily build up in the near stagnant groundwater. Eventually, gypsum crystals begin to grow from the concentrated liquid.

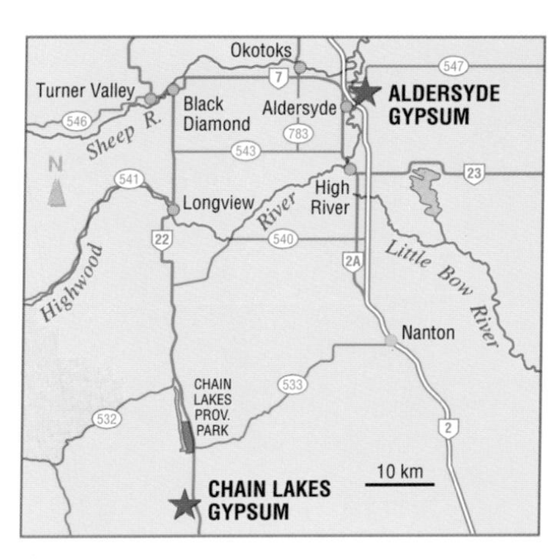


Another locality to search for gypsum crystals is one kilometre due east of the town of Aldersyde. The crystals are located in a ravine on the east bank of the Highwood River. There are outcrops of Paskapoo Formation sandstone within 100 metres of this gypsum occurrence. The crystals here, unlike those found at Chain Lakes, form as "rosettes." Rosettes consist of several single crystals that have grown together, with each crystal resembling a flower petal. Some rosettes reach seven centimetres in diameter! There is less clay in the soil here and sand grains have been incorporated into the rosettes, giving them a cloudy appearance.

*Gregory Baker – Provincial Museum of Alberta*

*This cluster of intergrown crystals of gypsum is called a "rosette" because of its similarity to a flower.*

Gypsum is one of the most common sulphate minerals. It is important in the manufacture of plaster of Paris, Portland cement, drywall, fertilizer, and even in casts for fractured limbs.

# CHAPTER 7: ALBERTA'S SOUTH

*Mt. Blakiston, Waterton Lakes National Park. Within the grey limestone beds is a black band of igneous rock called a sill. Note the sill is bordered by white marble where the limestone was cooked by the molten rock.*

*Ron Mussieux – Provincial Museum of Alberta*

# ALBERTA'S SOUTH

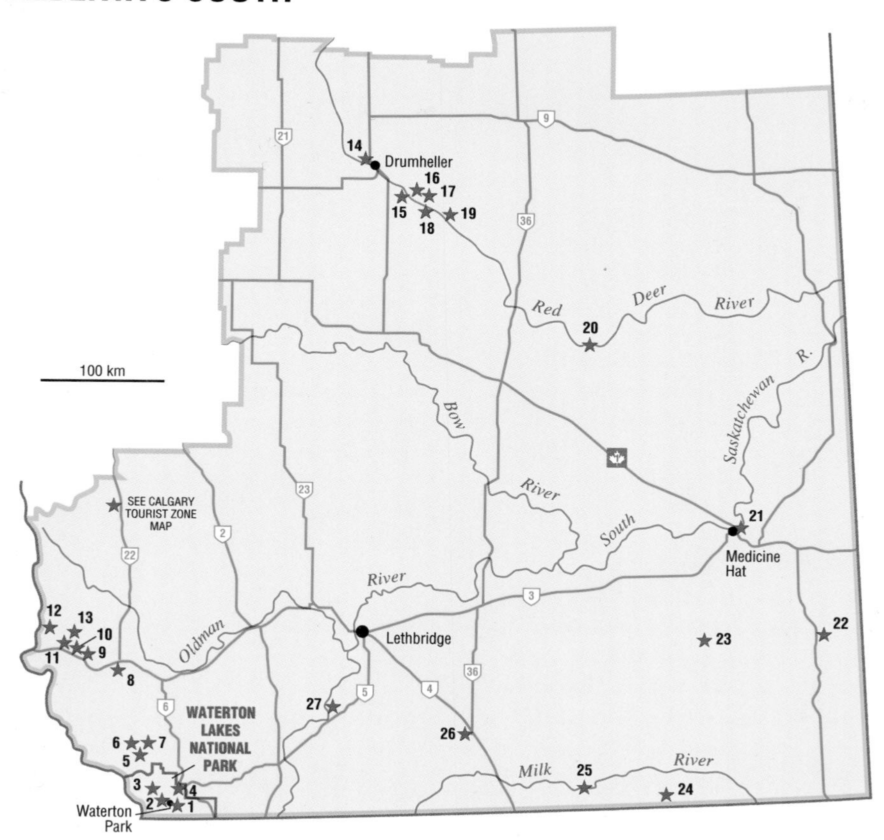


★ **1** Cameron Falls
★ **2** Oil City Discovery Well
★ **3** Red Rock Canyon
★ **4** Buffalo Paddock Hills
★ **5** Stromatolites
★ **6** Purcell Lava
★ **7** Waterton Sour Gas Plant
★ **8** Lundbreck Falls
★ **9** Bellevue
★ **10** Frank Slide
★ **11** Crowsnest Volcanics
★ **12** Crowsnest Mountain
★ **13** Grassy Mountain
★ **14** Royal Tyrrell Museum
★ **15** Drumheller Minerals
★ **16** Rosedale Burnt Shale
★ **17** Willow Creek
★ **18** Atlas Coal Mine
★ **19** Dorothy Bentonite
★ **20** Dinosaur Provincial Park
★ **21** Medalta Pottery
★ **22** Cypress Hills
★ **23** Red Rock Concretions
★ **24** Milk River Dykes
★ **25** Writing-on-Stone Provincial Park
★ **26** Devil's Coulee Dinosaur Egg Site
★ **27** Ammolite

# CAMERON FALLS: A Classic Hanging Valley

*Ron Mussieux – Provincial Museum of Alberta*

*The cascading water at this pretty waterfall outlines and increases the visibility of the arching rock beds.*

## HIGHLIGHTS

At this picturesque site, Cameron Creek plunges over 1.5-billion-year old rock, one of the oldest known outcrops in the Rocky Mountains. Cameron Falls is a textbook example of a waterfall formed over a hanging valley that was formed by glacial erosion during the last Ice Age. Over time, the creek has deposited its load of sediments into the valley and slowly constructed the broad, cone-shaped alluvial fan upon which the town of Waterton is built.

## THE STORY

Glaciers greatly alter the landscape. Everywhere in Waterton Lakes National Park there is evidence that massive glaciers once flowed through the area, leaving behind jagged, steep mountains, waterfalls, and U-shaped valleys. One characteristic feature of glaciated mountains is a hanging valley.

Over 12,000 years ago, an enormous glacier slowly scoured out the valley now containing Upper Waterton Lake. Smaller glaciers flowed into it from side valleys and the entire mass of ice crept northwards towards the prairies. When glaciers of different sizes meet, the valley containing the larger glacier is eroded wider and deeper than the valley with the smaller glacier. Thus, when the ice melts, the smaller side valley is left "hanging," or stranded, above the floor of the main valley. The streams flowing over these hanging valleys plunge down the steep walls of the main valley creat-

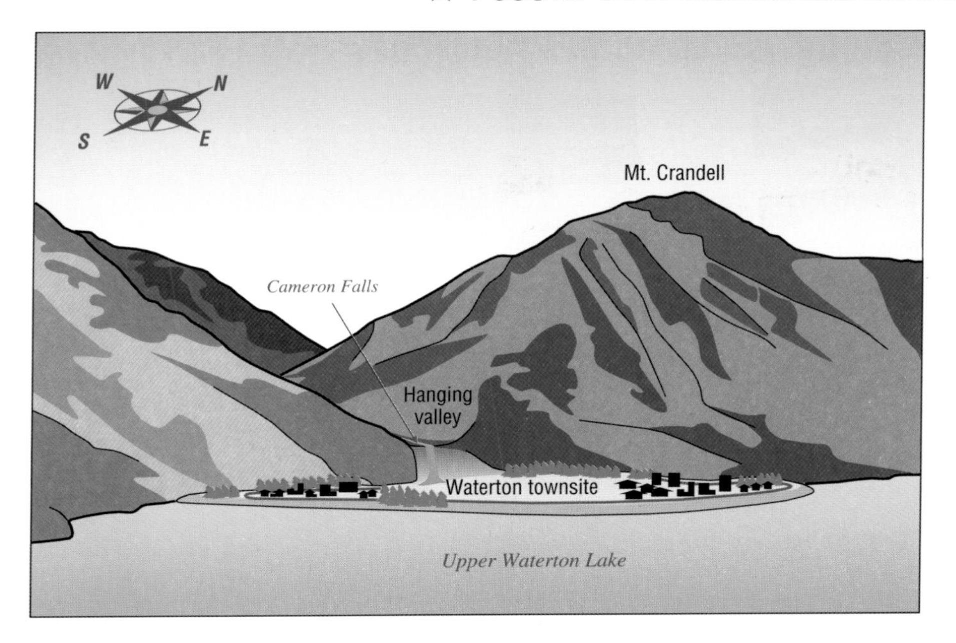


ing some of the most spectacular waterfalls in the world. In southern Alberta, Cameron Falls is a well known example of a hanging valley waterfall. Here, the water descends from the hanging Cameron Creek valley into the deeper, and larger Waterton valley. Other Alberta examples of waterfalls formed over hanging valleys are Bridal Veil Falls, Johnston Canyon, and Maligne Canyon.

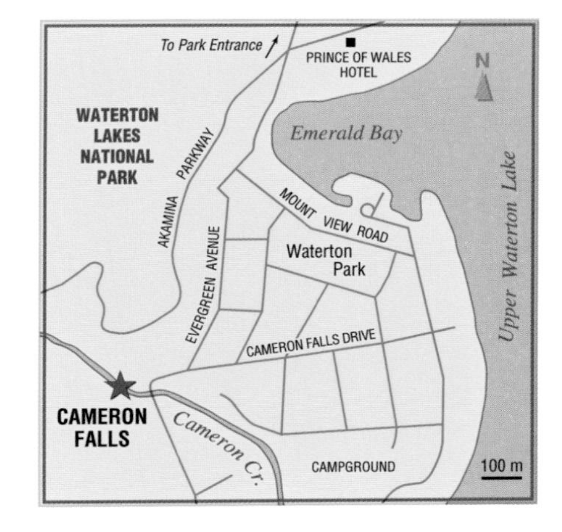


The rock layers forming Cameron Falls are limestone and dolostone that are folded into an arch, or anticline. These rocks are of the Precambrian Waterton Formation that formed 1.5 billion years ago as a mudflat of a warm, shallow ocean. Here, cyanobacteria thrived and their fossil remains, called stromatolites, are abundant.

## OIL CITY: Western Canada's First Oil Well

*Ron Mussieux – Provincial Museum of Alberta*

*Discovery Well monument. Note that the original drill stem is still "in place" at the bottom of the monument.*

### HIGHLIGHTS

Near Waterton townsite, along the Akamina Parkway, stands a monument to commemorate the site of Western Canada's first producing oil well and only the second in Canada. In the euphoria resulting from this 1902 oil find, the land was cleared and surveyed nearby and Oil City was built. Although the story of Oil City goes from dreams to bitter disappointment, it did foretell a potential oil industry in Canada, and particularly Alberta. Today, 1.3 kilometres southwest of the Discovery Well, a short path off the Parkway takes you down to the remains of the ill-fated town.

### THE STORY

The Stoney People (Nakoda First Nation) knew of oil seepages along Cameron Creek long before the Europeans came to Alberta. They used it for medicinal purposes, particularly to smear on open wounds on themselves and their animals. Kootenai Brown, a well-known Irish pioneer, was the first settler to find out the location of the oil from them. The story goes that he gave them a mixture of molasses and coal oil and asked if it tasted familiar. It did, and they led him to Oil Creek, now called Cameron Creek, where "black gold" oozed from the ground.

William Aldridge was the first to put

the oil to commercial use in the late 1800s. He soaked up the oil in gunnysacks and wrung it out into liquor jugs brought from Cardston saloons. He then sold his oil for a dollar per gallon to ranchers in Lethbridge, Fort Macleod, and Cardston for medicine and lubricating grease. His mining method was primitive, and it was still years before oil-starved prospectors came to the area.

In 1902, the Discovery Well was drilled by the Rocky Mountain Development Company. The well hit oil at only 312 metres and the boom was on. Entrepreneurs drilled wherever seepages showed, including Lineham Creek and near Cameron Falls where the government houses now stand. Drilling in this mountainous area, however, proved difficult. Equipment had to be winched by pulleys from tree to tree over the mountain passes. Many derricks blew over in the strong winds and pieces of equipment fell down wells (one story attributes this to deliberate sabotage by an eastern Canada oil company).

By 1904, the flow of oil in the discovery well had slowed to a trickle, and by the 1930s Oil City and all of the oil drilling equipment were abandoned. Some of this equipment can still be seen rusting in the creeks. Geologists believe that the oil that was first struck likely had seeped upward through fractures and along a fault plane. Seepages can still be found today along Cameron Creek and Lineham Creek.

NA 4089-4 – Glenbow Archives

*Original discovery well, drilled in 1902, on the right.*

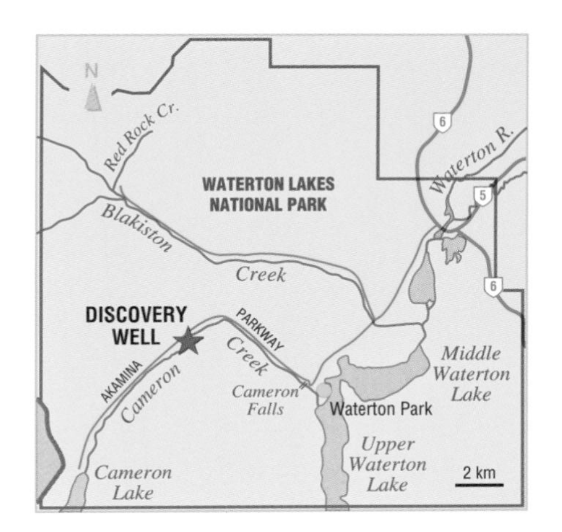

# THE ANCIENT IRON-RICH MUDS OF RED ROCK CANYON

*Ron Mussieux – Provincial Museum of Alberta*

*These bright red argillites of the Grinnell Formation were originally mud flats over one billion years ago.*

## THE STORY

The rocks that line Red Rock Canyon began as iron-rich mud on a tidal mudflat of an ancient ocean. The vivid red color of the rock layers was formed during times when the muds were exposed to the air. Similar to rusting, the iron was oxidized (oxygen was added) and formed the red mineral hematite. Only a small amount of hematite in a rock (3 per cent) is necessary to produce such a brilliant red color. The green layers were formed when the mud was not exposed to the air. With no air, the iron was reduced, which means oxygen was removed, allowing the green mineral chlorite to form. This might have occurred when water levels were higher, or there was increased rainfall, or even if the chemistry of the water changed.

Over millions of years the mud was buried, subjected to heat and pressure beneath the earth's surface, and turned into a slightly metamorphosed rock called argillite. The same red argillites of the canyon can also be seen in many of the mountain peaks in the park, such as Ruby Ridge, Vimy Peak, and Mt. Galwey.

## HIGHLIGHTS

The walls of this canyon are streaked red and green by different minerals in the rocks, making this a popular site in Waterton Lakes National Park. These rocks are some of the most colorful and oldest rock formations in the Rocky Mountains, at over one billion years of age. Ten thousand years of erosion by cascading water has cut through the rock layers to form the 23-metre deep-canyon. A short interpretive trail winds along and into the canyon.

Layers of fine white sandstone may also be seen in the canyon. These were deposited during violent storms by flooding rivers that drained from the mainland to the east. Occasionally storms tore up the mud and redeposited it as angular red chunks in the sand. The resulting rock is called mudchip breccia.

Red Rock Canyon contains many excellent examples of sedimentary features. Mud cracks and salt crystal casts, seen on the east side of the canyon, formed as the wet mud dried and cracked during low tides. Ripple marks are evidence of gentle currents and waves on the mudflats. Impressions of large raindrops are also preserved in some rock layers, although these are more difficult to find. These features are all evidence of an ancient ocean shoreline environment, with alternating wet and dry conditions.

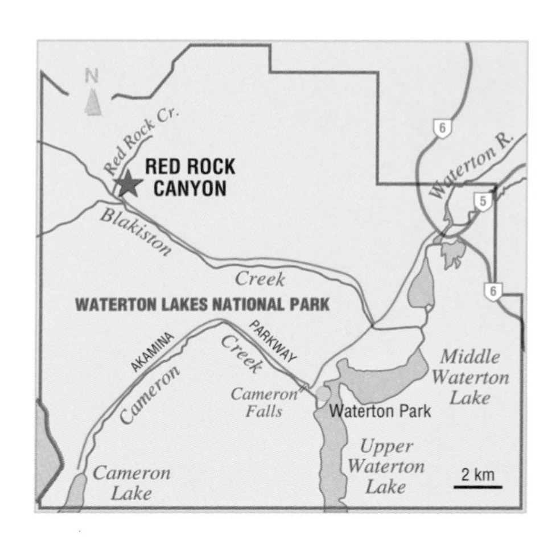

Ron Mussieux – Provincial Museum of Alberta

*These shallow mud flats frequently dried and developed large cracks.*

# THE GLACIAL ORIGIN OF THE RIDGES AND HILLS OF THE WATERTON BUFFALO PADDOCK

*Ron Mussieux – Provincial Museum of Alberta*

*Linear ridge, called an esker.*

## HIGHLIGHTS

The present landscape of Waterton Lakes National Park, like that of most of Canada, was modified by the action of glaciers in Earth's recent past. Glaciers both erode and deposit sediments, drastically changing the original landscape. Although glacial erosional features, such as U-shaped valleys and polished bedrock, are quite obvious to the public, the depositional features associated with glaciers are often overlooked. The Buffalo Paddock, a rolling grassland on the eastern boundary of the Park, preserves some of Alberta's best and most accessible depositional features. Sands and gravels carried by water from melting glaciers were deposited along the edges of melting ice to form long sinuous ridges, called eskers, and small, conical hills, called kames.

## THE STORY

The term esker comes from the Irish word "eiscir" meaning ridge. Geologists define an esker as a long, narrow, sinuous, steep-sided ridge composed of irregularly layered sand and gravel. It was deposited by a stream that ran on top of, under, or at the margin of a glacier, and, as the glacier melted away, it left behind these branching and often discontinuous ridges. The biggest esker near the paddock is over three kilometres long, from 10 to 60 metres wide, and is nearly 18 metres high — not an insignificant pile of gravel! Of historical interest, it is believed that this esker was used by Aboriginal People over 8000 years ago as a shelter from the fierce Chinook winds and as a bison-trapping drive lane. Eskers are a particularly common landform in Canada's far north where they can be as long as 500 kilo-

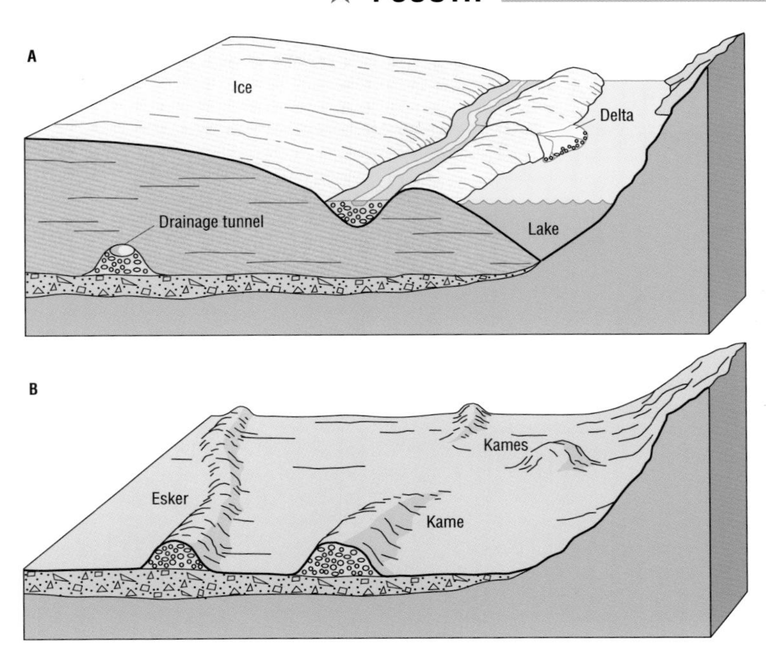


*Modified from Misc. Report 26 (1976) courtesy Geological Survey of Canada*

metres and up to 100 metres high. These northern eskers were built by massive continental glaciers and became important to prospectors because they were used to locate mineral deposits including diamonds. They will eventually lead to the opening of Canada's first diamond mine. The eskers in Waterton Lakes National Park are not nearly as large as their northern counterparts, however, because they were only built by small valley glaciers.

Kames are similar to eskers in that they are composed of sand and gravel, but they are hills instead of ridges. They were deposited in three different areas on a melting glacier: in a delta of a stream flowing from the base of the ice; in a hole on the surface of the ice where a stream trickles down; or as a pond deposit on the surface or at the margin of the glacier. A few kames can be seen northeast of the Buffalo Paddock and the lookout is located on top of one. A special type of kame, called a kame terrace, sits below Waterton townsite. These long, flat-topped hills were formed as meltwater flowed between the margin of a glacier and valley wall and deposited mounds of sand and gravel.

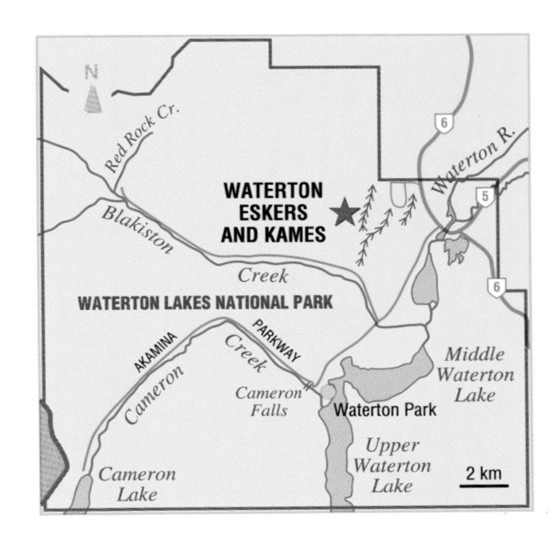

# SOUTH DRYWOOD CREEK STROMATOLITES:
## Alberta's Oldest Fossils

*Ron Mussieux – Provincial Museum of Alberta*

*Limestone block with large stromatolite heads, some over 30 centimetres across.*

## HIGHLIGHTS

Stromatolites are the fossil deposits of microscopic organisms, called cyanobacteria. They have the distinction of being the oldest fossils in Alberta, at over a billion years old. Their presence in rocks helps mark the locations of ancient coastlines in Alberta.

## THE STORY

Before most other life forms existed, cyanobacteria thrived and flourished in shallow coastal pools along most of the earth's shorelines. Growing in enormous colonies that resembled giant cabbage heads, each colony was like a mat which often covered an area 10-20 centimetres across. They grew and died beside and on top of each other, gradually forming massive reef-like accumulations with heights up to 10 metres. As they grew, they deposited layers of calcium carbonate ($CaCO_3$), producing a concentric, laminated structure. It is these $CaCO_3$ layers that we find, not the soft bodies which decomposed after they died. This makes stromatolites a peculiar kind of "trace" fossil because they do not represent the actual remains of organisms, but rather the growth and life processes of this primitive type of bacterium.

Cyanobacteria were photosynthetic and at their peak were a major source of Earth's oxygen. The increased oxygen levels in the atmosphere allowed for higher life forms to develop and evolve on our early planet. They made up the majority of the world's reefs and became havens for other plants and eventually even for marine animals. The

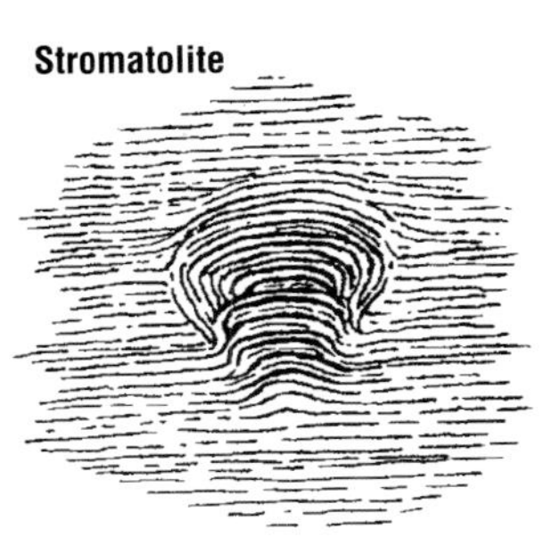

Ron Mussieux – Provincial Museum of Alberta

*Stromatolite colonies often grew together to form massive reefs. Note the concentric growth structure.*

oldest stromatolites dated so far are from northern Western Australia and Zimbabwe, and are about 3.5 billion years old. Cyanobacteria still survive today although they are not as wide-ranging as they once were. There are still some growing in the shallow tidal waters along the coastlines of Western Australia and the Persian Gulf.

These fossils are actually quite common and are found in many rock formations in the Rocky Mountains. Unfortunately though, their rather vague, nondescript appearance makes them difficult to recognize. One easily accessible locality is along South Drywood Creek Road, also called Bovin Lake Road. There is a good gravel road up to Wellsite 7, but beyond this it becomes a rough 4-wheel drive vehicle trail. The main stromatolite reef outcrop is 2.8 kilometres past this wellsite, although you will find numerous fall blocks of these fossils along the side of the ridge. The road crosses the outcrop and the reef forms a steep cliff on the northwest side of the valley. At this location, the fossil heads are quite large, up to 30 centimetres in diameter, and have grown together to form the massive reef structure. On some rock faces, you can see the laminated structure and the complex intergrowth of colonies.

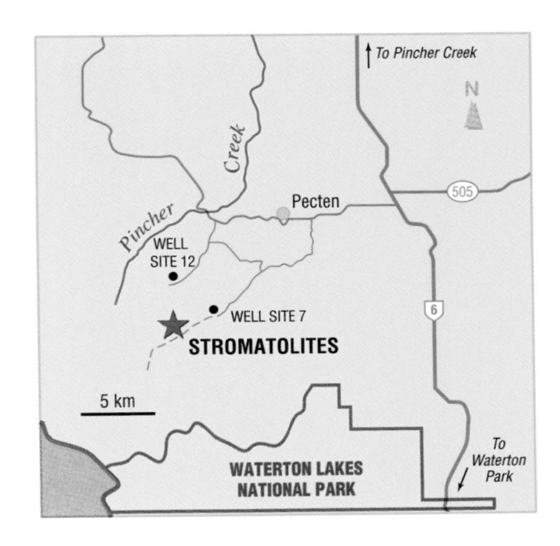

# PURCELL LAVAS: An Ancient Alberta Lava Flow Chilled by the Sea

*Ron Mussieux – Provincial Museum of Alberta*

*Waterfall on Drywood Creek over the ancient Purcell lava flow.*

## HIGHLIGHTS

Volcanic igneous rocks, formed by the eruption of molten rock onto the earth's surface, are rare in Alberta. There are, however, two kinds of volcanic rocks in southwest Alberta — the Crowsnest Volcanics (see page 206), which is evidence of an explosive eruption similar to Mt. St. Helens, and the Purcell Lavas, which were similar to the fluid, gentle lava flows in Hawaii today. The Purcell lavas are found in both Waterton Lakes National Park, Alberta, and Glacier National Park, Montana. One of the best sites outside the parks to see this lava is Drywood Creek which is accessible by road through the Shell Waterton Gas Field.

## THE STORY

Lava is molten rock that has flowed onto the earth's surface and hardened. Drywood Creek, north of Waterton Lakes National Park, contains thick flows of dark-colored basaltic lava, called the Purcell Lavas. Basalt is a common type of volcanic rock, rich in magnesium and iron. It is often full of holes, once air bubbles in the lava, which have since filled in with calcite and quartz crystals. Occasionally, basalt will even contain sedimentary rock fragments that were incorporated as the lava flowed over them.

An excellent outcrop of Purcell lavas occurs as the ledge of a waterfall on Drywood Creek, about 400 metres upstream of the Shell 12 Waterton Well. This basalt has been altered on the surface to dark purple and green and at the base of the outcrop you can see peculiar bulbous shapes, called "pillow" lava. Pillow lava forms only in an underwater environment. As lava oozes into the water, it flows as tubes which continually divide, feeding ever-branching pillow lobes. These lobes have a glassy crust because the lava was quickly cooled by the cold water.

*Don Taylor – Provincial Museum of Alberta*

*Well developed pillow lava on South Drywood Creek trail.*

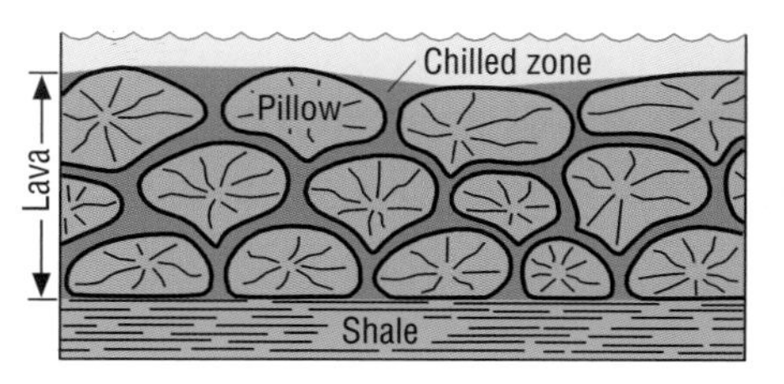


Another outcrop of this lava is located in South Drywood Creek, where a 4-wheel drive vehicle trail runs over the basalt about 3.5 kilometres upstream of Shell 7 Waterton Well. Here, the basalt is around 60 metres thick and forms steep cliffs.

In Waterton Lakes National Park there are outcrops of igneous rock which forms sills and dykes. These were formed when molten rock, or magma, flowed up into subsurface cracks and hardened without reaching the earth's surface. Where the magma cuts across bedding, we call it a dyke; where it flowed along fractures parallel to bedding we call it a sill. Millions of years of erosion have exposed them at the surface. Sills and dykes look different from Purcell lava because they lack pillow structures and because the rocks on both sides of them are baked by the molten rock. On the road to Red Rock Canyon in Waterton Park you can see white marble, or "baked limestone," above and below a sill on the southeast face of Mt. Blakiston (see photo page 184). Dating techniques indicate that the lava, sills and dykes are all about 1.1 billion years old.

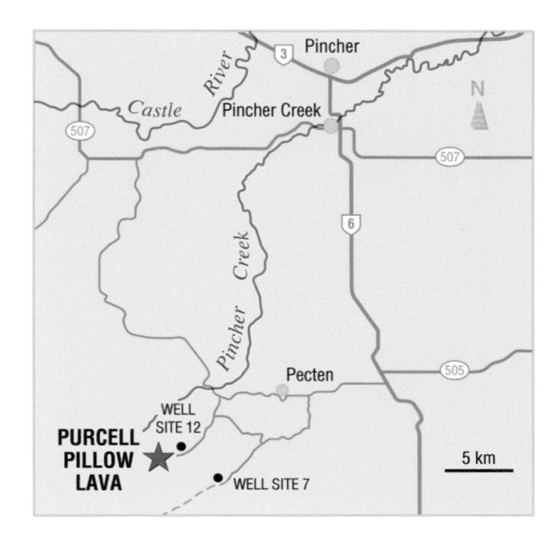

# SULPHUR: Waterton Sour Gas Plant

*Ron Mussieux – Provincial Museum of Alberta*

*Shell Oil Company's Pecten Plant, Waterton Sour Gas Field.*

## HIGHLIGHTS

The yellow mounds of sulphur found at the Shell gas plant near Waterton Lakes National Park are a strange and distinctive sight. They offer a stark contrast to the snow capped mountains and rolling green foothills of Alberta. The sulphur found here is a co-product of Alberta's billion dollar natural gas industry. The sulphur mounds grow and shrink depending on the market, but production of the yellow mineral continues non-stop as long as the natural gas wells continue to operate.

## THE STORY

Alberta is one of the world's largest producers of sulphur. Here, it is found in natural gas fields, not in solid form but rather as hydrogen sulphide gas. Natural gas that contains hydrogen sulphide is referred to as "sour" gas, whereas sulphur-free gas is called "sweet" gas. Most sour gas fields contain up to 20 per cent hydrogen sulphide but concentrations as high as 90 per cent have been recorded in Alberta's Foothills. Hydrogen sulphide is a poisonous gas with a rotten-egg odor. The odor can be detected by humans at concentrations of 0.1 parts per million (ppm) and is rapidly fatal at concentrations over 600 ppm. This is the same odor that can be smelled at Miette Hot Springs and Cold Sulphur Springs, but the concentrations of hydrogen sulphide at these locations are not hazardous. Hydrogen sulphide must be removed from the sour gas at extraction plants, such as the two near Waterton where it is transformed into sulphur. The extraction process can remove as much as 99.9 per cent of the hydrogen sulphide in the gas. The little that remains is burned off in an incinerator to form sulphur dioxide which is then released into the atmosphere through one of the towering smokestacks. The sulphur was once considered a waste product of the industry and as the piles grew it was viewed as a nuisance and an environmental hazard. Alberta's sulphur is now used in the fertilizer industry, rubber products, explosives, and petroleum refining.

Sour gas fields are present throughout Alberta, except for the far northeast. Most lie in a 80-kilometre-wide stretch of the Foothills that runs the length of the province. The Waterton field is tapping gas from rocks which are between 400 and 320 million years old and that lie beneath much older rocks. In order to get down to these younger reservoirs, wells had to be drilled through thrust-faulted rocks that were over one billion years old. The original sulphur in the gas is believed to have been produced as a by-products of bacterial action.

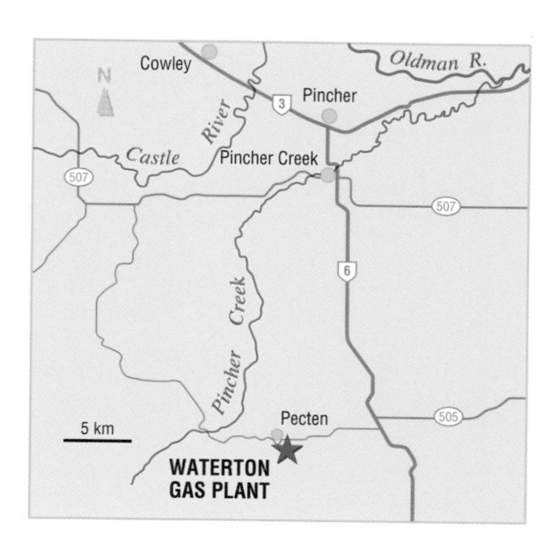

*Ron Mussieux – Provincial Museum of Alberta*

*A sulphur stock pile. Notice the metal frame that holds the liquid sulphur until it hardens into a solid.*

# LUNDBRECK FALLS AND THE EASTERN END OF CROWSNEST PASS

Julie Hrapko – Provincial Museum of Alberta

*Spring runoff cascades over Lundbreck Falls.*

## HIGHLIGHTS

At the eastern end of the Crowsnest Pass, the Crowsnest River plunges 17 metres over a sandstone cliff to form Lundbreck Falls. Located 12 kilometres west of Pincher Creek, these falls are near the eastern margin of the Foothills. Due to the magnificent exposures of rocks in this region, plus the abundant seismic testing and petroleum drilling, it is a principal reference area for the study of Foothill structure. The Foothills form a belt of low rolling hills, grasslands, and steep-walled valleys that is bounded on the west by the Front Ranges of the Rocky Mountains and on the east by the flat-lying rocks of the Porcupine Hills, part of the Interior Plains.

## THE STORY

Lundbreck Falls is formed where the Crowsnest River plunges over an erosion-resistant sandstone cliff of the Belly River Formation, deposited around 75 million years ago. This formation was deposited in a broad delta that bordered the western shoreline of the Late Cretaceous inland sea that covered much of Alberta. During this time, rivers flowing eastwards from the rising mountains in the west brought great quantities of sediment, particularly sand, and dumped them on the many deltas, such as the Belly River Delta. The cliff that forms Lundbreck Falls contains numerous sedimentary structures, such as ripple marks and worm burrow traces, which indicate a shallow shoreline environment. The Belly River sandstone is an extensive, wedge-shaped formation that underlies much of Alberta. Petroleum geologists often use it as a reference layer in seismic printouts to determine the relative position and depth of other known petroleum-bearing formations.

Looking towards the west from this site, you can see the folded and faulted Paleozoic limestones of the Front Ranges. The rocks of the Foothills are also heavily folded and faulted, but are younger, of Mesozoic age, and consist of sandstones and shales that form northwest-trending ridges. Because these are more susceptible to erosion than the limestones of the Front Ranges, the original height of the Foothills has been eroded away more quickly, leaving behind an area of low forested hills and grasslands.

Ron Mussieux – Provincial Museum of Alb

*Steep dipping Cretaceous sandstone of the Foothills, just east of Lundbreck Falls.*

As you travel eastwards across the Foothills, the topography becomes flatter and less distinct until it finally grades into the Plains. North of Lundbreck Falls lie the Porcupine Hills; these are the western boundary of the Plains.

Also of interest in this area is the excavation of "Black Beauty" in 1982. Black Beauty is a beautifully preserved, rich-black *Tyrannosaurus rex* that was discovered about 5.5 kilometres east of Lundbreck along the Crowsnest River. The dinosaur was prepared for display at the Royal Tyrrell Museum of Palaeontology in Drumheller.

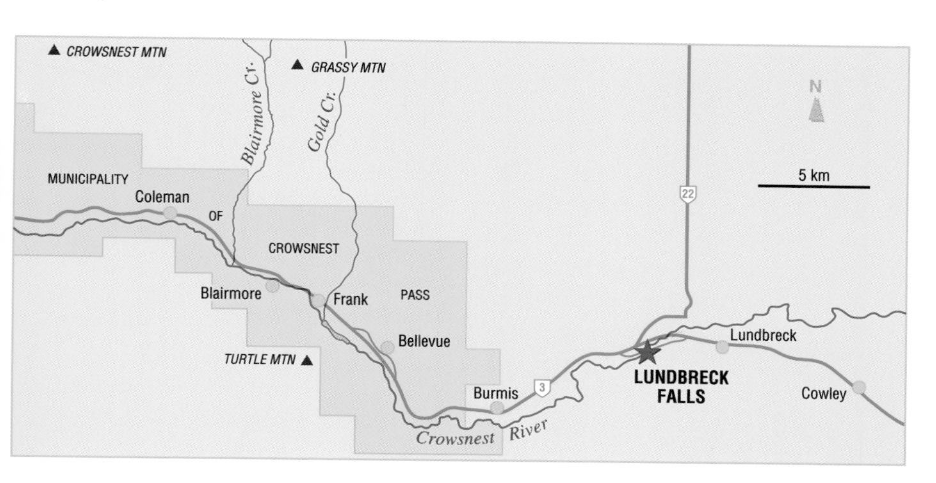

## BELLEVUE: Underground Coal Mining in the Crowsnest Pass

PA 1981-1 – Provincial Archives of Alberta

*Bellevue coal mine in full operation, 1951.*

### HIGHLIGHTS

The Crowsnest Pass encompasses a number of once-bustling coal-mining towns, such as Bellevue, Frank, Coleman, and Hillcrest. The tumultuous history of these communities is one of strikes and tragedies, and in 1988 the Pass became Alberta's first designated Ecomuseum and Historic District. The Bellevue Mine, which closed in 1962, is the only mine in this area that has been partially restored and opened to the public; an excellent underground mine tour gives a realistic glimpse into a coal miner's life.

### THE STORY

Even though Crowsnest Pass was recognized as containing one of Canada's greatest concentrations of coal, it was not until 1898 when the Canadian Pacific Railway opened a line through the Pass that it became a major coal-producing region. The Bellevue town and mine, named for its beautiful view high in the mountains, was opened in 1903 by Western Canadian Collieries Ltd., based in Lille, France. The thickest coal seam was located right beside the main CPR tracks, and Bellevue became the first coal town in southern Alberta to provide fuel for the railways in this steam era. Although marketing the coal was easy, the extraction of it was a problem. Even though the seam was 3.5 metres thick, it was steeply dipping and

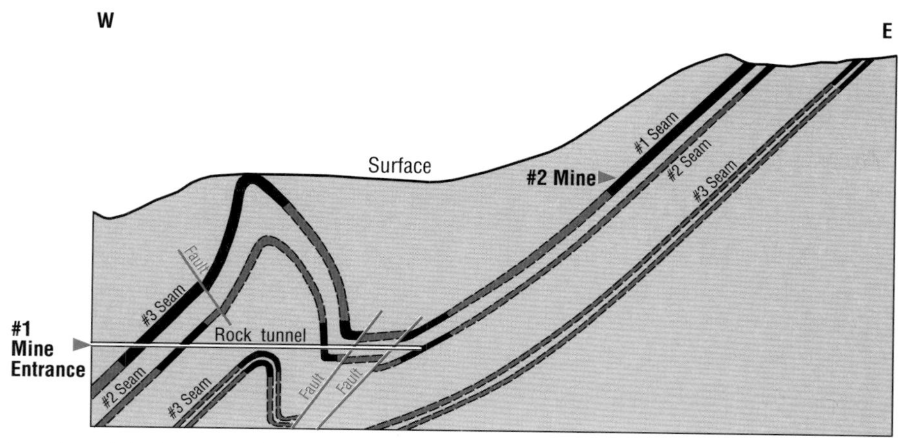


*A section of the Bellevue Mine, published in 1913, showing the folded and faulted nature of the coal seams.*

usually discontinuous. As well, it contained pockets of explosive methane gas. The operation ran safely until December 9, 1910, when an underground explosion rocked the mine and demolished the ventilation fan. Likely, rock falls from the roof ignited some methane gas which then produced a coal dust explosion. After the oxygen was burned from the air, high concentrations of carbon monoxide and carbon dioxide were left. These gases killed 30 miners. This was only the first of several accidents in the Pass mines, which led to new safety measures in mines in both Alberta and British Columbia.

The mine reopened after this disaster and, by the mid-1920s, had reached its maximum production of 2300 tonnes per day of high rank bituminous coal. The mine extended 850 metres into the mountain and coal was extracted on nine different levels of workings. Various methods were used to load and transport the coal to the surface but the actual extraction of the coal from a seam was done mostly by hand. This was quite a feat when we consider that this mine produced 1,364,000 tonnes of coal in its 60-year history!

During its peak years, at least half the coal was made into coke to fuel smelters in British Columbia and the northwest United States. However, the increasing demand for oil and gas led to a decline in the coal market and many of the coal mines closed. Equipment from nearby coal mines is on display in Bellevue and the interpretative building has great educational value. For more information contact the Frank Slide Interpretive Centre.

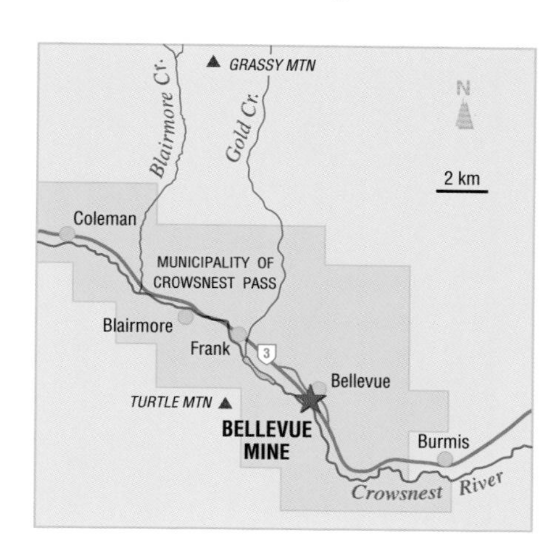

# FRANK SLIDE: A Crowsnest Pass Tragedy

*C. Wallis – Natural Resources Service*

## HIGHLIGHTS

On April 29, 1903, at 4:10 a.m., 30 million cubic metres of limestone hurtled down the east face of Turtle Mountain. Boulders piled up to 30 metres deep cover more than 2.5 square kilometres of the Crowsnest River valley in awesome testimony to the destructive power of the Frank Slide. In only 100 seconds, boulders were spread across the 1000 metre wide valley and 120 metres up the opposite valley wall. Part of the town of Frank was buried, the rail line was barricaded, the entrance to the coal mine was temporarily sealed, and over 70 people were killed. Because of the potential for further slope failure, Turtle Mountain is monitored for any perceptible movement. The Frank Slide Interpretive Centre sits high above the rubble across the valley from the mountain, and a 1.5-kilometre trail leads down to the slide itself. Frank Slide is a designated Provincial Historical Site.

## THE STORY

This rockslide has become the classic example of mass movement because it is one of the largest slides to have eyewitness descriptions as well as numerous detailed geological reports. Although we will never know the final trigger that set the slide in motion, we do know that the fundamental structure of Turtle Mountain was unstable from the onset.

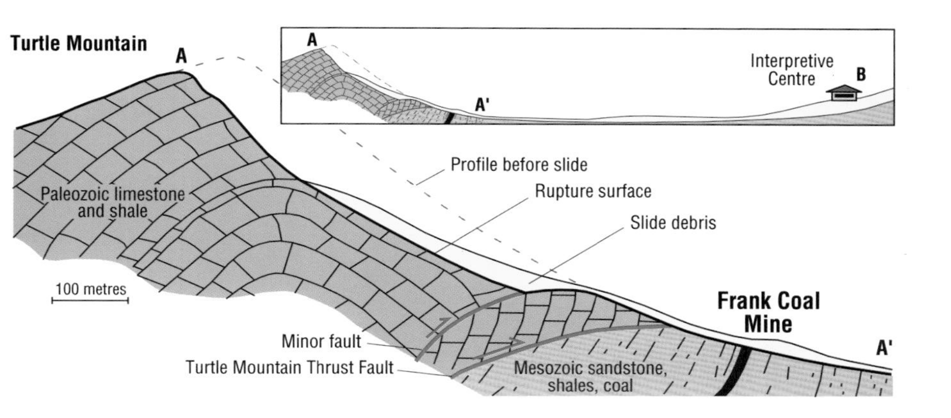


The rock layers of Turtle Mountain are folded into a large anticline with a steeply dipping east limb that lies on a fault. Many miners believe that mining and blasting at the base of the mountain weakened the mountain. Before the slide, they had commented on the cracking of the wood supports in the mine. As well, prior to the slide, cracks had formed in the exposed bedrock at the mountain's summit. Local weather conditions just before the slide indicate that it was hot and so snow would have melted and filled these cracks with water. The temperature turned cold on the night of April 28 and the water likely froze and expanded, wedging the cracks apart. This, then, may have been the final strain on an already unbalanced mountain. Whatever the reason, once the rock mass was in motion, it slid down along bedding planes in the limestone and then followed the plane of the fault at the foot of the mountain.

One of the more popular theories to explain the movement of such an enormous wedge of rock is called "air-layer lubrication." This means that the entire rock mass rode on top of a trapped layer of air to its present location. Although this does not explain everything, it does account for the rapidity of the event, the vast lateral extent of the boulders, the lack of rocks at the immediate foot of the mountain, and the steepness of the margins of the slide. Many eyewitnesses have said that they heard or felt blasts of air escaping from around the fringes of the slide "resembling that of steam escaping under high pressure."

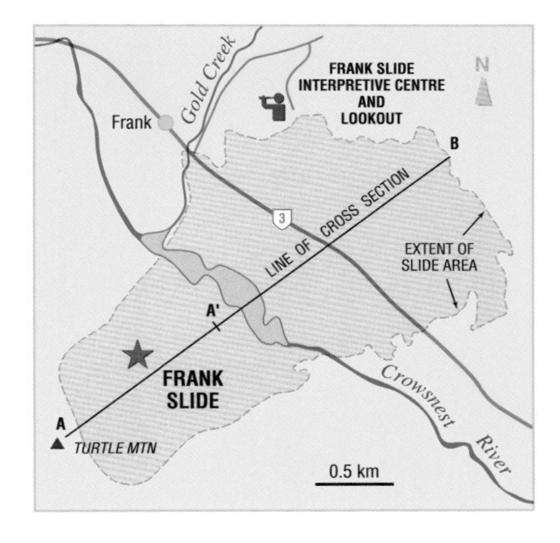

# CROWSNEST VOLCANICS: Alberta's Catastrophic Volcanic Eruption

Ron Mussieux – Provincial Museum of Alberta

*Large block of Crowsnest Volcanics made up of rock fragments that were ejected during the eruption.*

Ron Mussieux – Provincial Museum of Alberta

*Highway 3 outcrop of the 100- to 96-million-year old Crowsnest Volcanics in the foreground and the unrelated Crowsnest Mountain composed of much older sedimentary rocks in the background.*

## HIGHLIGHTS

One of the few exposures of volcanic rocks in the Rocky Mountains lies in Crowsnest Pass. These rocks, called the Crowsnest Volcanics, are evidence of a set of devastating, explosive eruptions dating from the time of the dinosaurs. The Crowsnest Volcanics are one of the province's most interesting and rewarding mineral collecting sites and they are well exposed along parts of Highway 3. Fortunately, this area is not volcanically active today.

## THE STORY

About 100 to 96 million years ago, between the rising Rocky Mountains to the west and the shore of Alberta's inland sea to the east, volcanoes exploded with a violence easily rivalling the catastrophic 1980 eruption of Mount St. Helens. Unlike the liquid lavas that are erupted from many volcanoes, the Crowsnest Volcanics erupted explosively, rupturing the magma into a torrential hail of solid rock debris, ash, and gases. The Volcanics consist mainly of airborne debris deposited directly from the explosive event and debris laid down in water. The original volcanic vent may be just north of Coleman, where the volcanic rocks are over 425 metres thick. Although the deposits thin in every direction away from this spot, no discernible remnant of a volcano can now be seen. Its true location remains a mystery.

The explosive nature of the Crowsnest eruptions resulted in debris ranging from fine ash to blocks several metres across. Volcanic rocks made up of very angular rock fragments are the most common deposits. Particularly noticeable are the bedded layers, where the

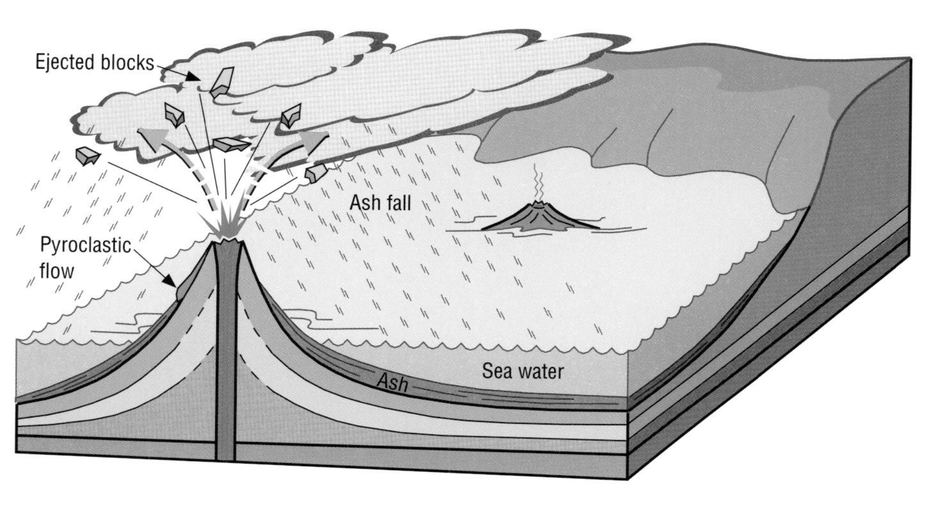


volcanic debris was reworked by water before it was deposited.

There is a great variety of rock types in the Crowsnest Volcanics. One type is made up mainly of the mineral analcime and the rock is named analcimite. This rock was once named blairmorite, after the town of Blairmore. Geologists now name igneous rocks after the major minerals forming them, instead of the locality where they are found. Other rock types contain crystals of pink sanidine feldspar and small black melanite garnets.

The most accessible Crowsnest Volcanics outcrop from which to collect minerals is exposed along Highway 3 at the west end of Coleman. Here, you can find lustrous black melanite garnets, ranging in length from two to five millimetres, and pink sanidine crystals. Large analcime crystals are best collected along the gas pipeline south of Crowsnest River, near Coleman.

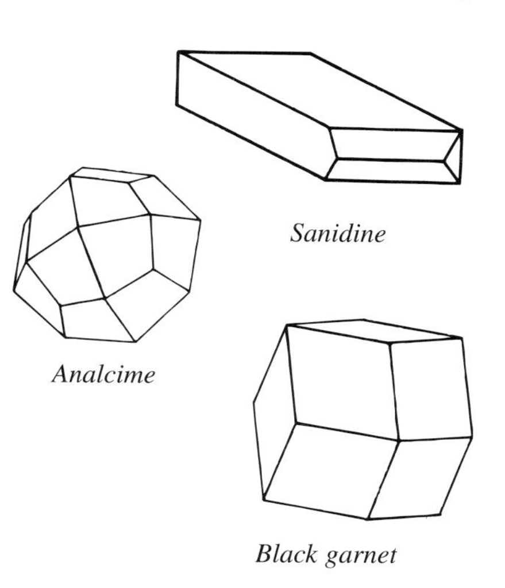


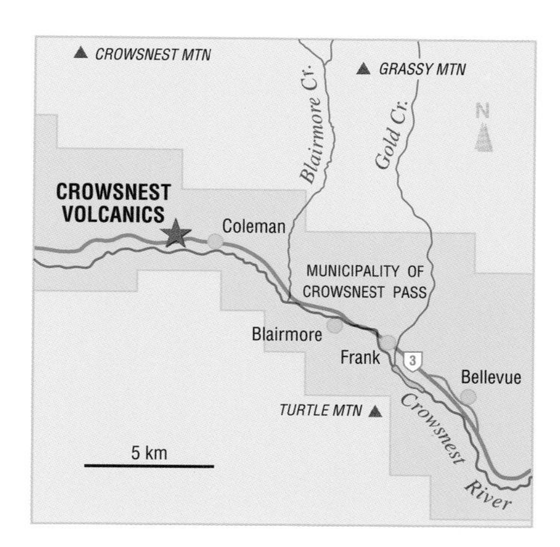

# CROWSNEST MOUNTAIN: The Mountain That Was Left Behind

*Ron Mussieux – Provincial Museum of Alberta*

*The isolated Crowsnest Mountain was once connected to the High Rock Range seen on the left of the photo.*

## HIGHLIGHTS

The magnificent Crowsnest Mountain guards the north side of the Crowsnest Pass, and has become the symbol of this famous pass. This mountain is a classic example of a klippe, the erosional remnant of a once-continuous mass of rock. To hike up this mountain, take the Allison Creek Road to the marked trailhead where there is a nearby parking spot. The summit provides a spectacular view into British Columbia to the west and of the Porcupine Hills to the east.

## THE STORY

Crowsnest Mountain is striking because it stands isolated from the mountains around it. It is interesting geologically because it is composed of the same rock formations as the High Rock Range to the west and, in fact, the two mountains were once connected. They were part of the same enormous slab of rock, called a thrust sheet, that was shoved eastwards over younger rocks during the building of the Rocky Mountains. Movement of the thrust sheet was along the Lewis Thrust Fault, one of the major faults in both the Alberta and Montana Rockies. It can be traced from Glacier National Park to the Crowsnest Pass and into the Central Rockies. Looking north at Crowsnest Mountain and High Rock Range, this

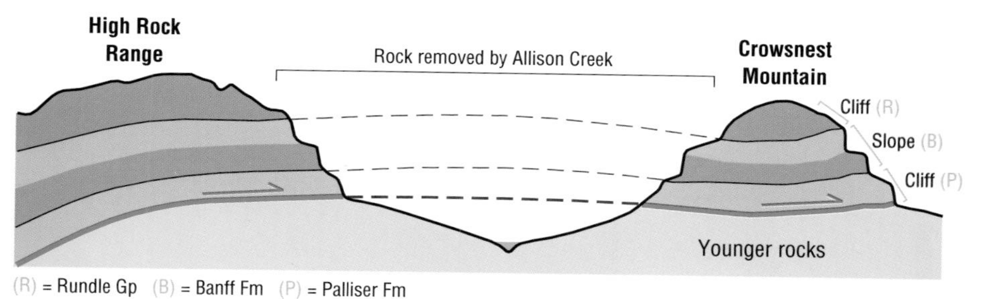


fault is almost horizontal and can be seen just above the tree line. Above the fault are limestones and shales of Paleozoic age, and below it are younger sandstones and shales of the Cretaceous Period. This is the reverse of the normal sequence of rock layers, where layers become younger as you move up a column of rock. Older rocks thrust on top of younger rocks, and separated by a thrust fault, is the classic signature of the Canadian Rockies.

Millions of years of erosion of this thrust sheet and the formation of the Allison Creek valley, created Crowsnest Mountain and left it detached from the main mass of the Rocky Mountains. Klippes are not uncommon in the mountains and, farther south, Chief Mountain in Montana is another excellent example.

Crowsnest Mountain is composed of the same rock formations that are seen in many other Front Range Mountains throughout the Rockies. These formations are an alternating sequence of limestone-shale-limestone that produces a distinctive cliff-slope-cliff pattern. The top of the mountain is the cliff-forming Mississippian Rundle limestone; the slopes below this are the slightly older soft shales of the Banff Formation (Mississippian Period); and the massive limestone cliff just above the tree line is the Devonian Palliser Formation.

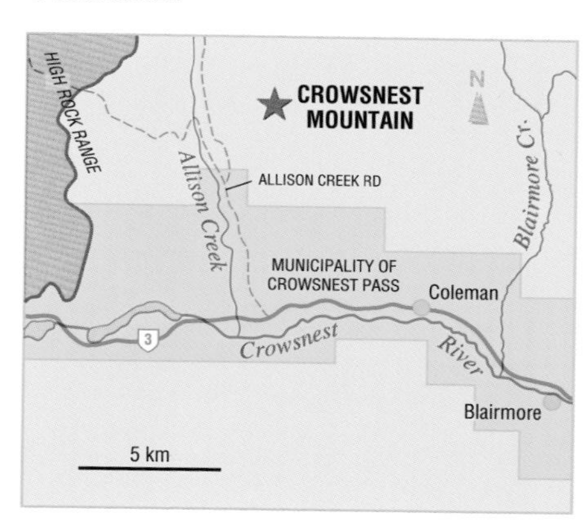

# GRASSY MOUNTAIN: The "Big Show"

*Ron Mussieux – Provincial Museum of Alberta*

*Structurally thickened coal seam called the "Big Show" or the "Pod," Grassy Mountain.*

## HIGHLIGHTS

Crowsnest Pass is famous for its extensive coal deposits and storied mining history. Located just north of the Pass, Grassy Mountain is a geological curiosity. An abandoned open pit coal mine on the south face of Grassy Mountain reveals a large deformed 25-metre thick coal seam, called the "Big Show" or the "Pod." As the old mining road is cut by erosion half-way up to the open pit, be prepared for a 20-minute hike up the hillside to the outcrop. The coal face at the site is steep and unstable, so please be careful not to disturb it.

## THE STORY

Grassy Mountain has an elevation of 2065 metres and is composed of three thrust sheets separated by two west-tilting thrust faults: the Turtle Mountain fault exposed just west of the open pit and the McConnell thrust fault found farther west. The most important layers are the 140-metre thick deposits of the Cretaceous Kootenay Group which contains the four major coal seams. Of these seams, the No. 2 coal seam is the thickest, closest to the surface, and the most important. These coal-bearing sediments are between 160 and 135 million years old and were deposited in deltas and adjacent coastal plains. The coal was originally deposited as peat in nearby swamps. Ripple marks, which were created by the running water in an ancient channel, can be seen at locations east of the open pit. Also, the roots of

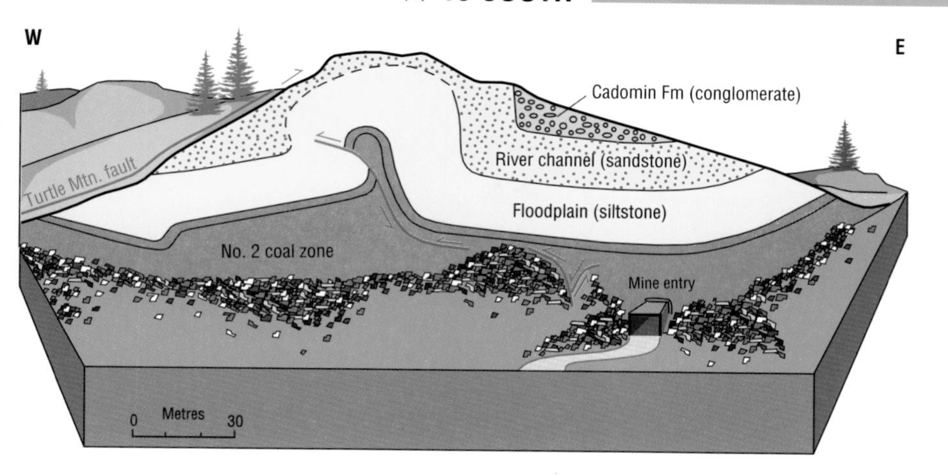


large trees are often preserved. These root systems, some having a radius of 1.5 metres, were originally horizontal in their swamp environment but have since been tilted until they stand in a near-vertical position.

The intriguing feature of the mountain is the thickening of the coal seams that took place during mountain building. The most noticeable is the coal "pod" of seam No. 2 which is exposed in the core of the anticline in the open pit. This highly deformed seam was thickened from 6.5 metres to an astonishing 25 metres because faulting caused an overlapping of the coal layers. In addition, the soft, plastic-like coal was squeezed into the centre of the anticline which further thickened the sequence. This thickening is what made the coal accessible and exploitable. However, the faulting and folding caused extensive fracturing and pulverizing of the coal seam. This made the extraction of the coal expensive and eventually forced closure of the mine.

Coal was found at Grassy Mountain near the turn of the century, and Western Canadian Collieries Ltd. began mining it in 1904. Over 3.5 million tonnes of bituminous coal were extracted from the mountain before market failure forced WCC to close operations in 1957. Since then, over 370 holes have been drilled in the mountain giving us a wealth of knowledge about the geological structure.

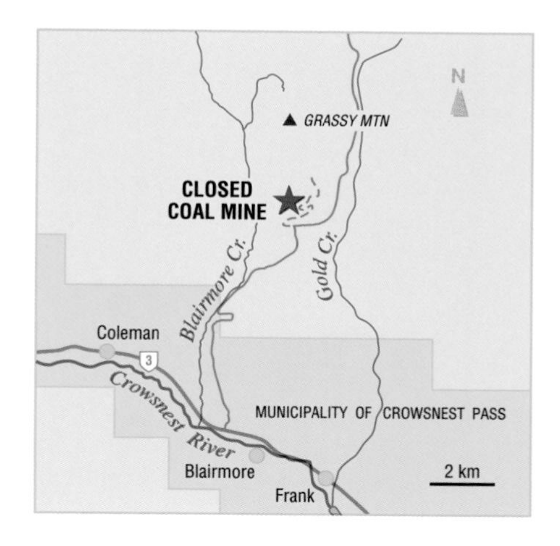

# ROYAL TYRRELL MUSEUM OF PALAEONTOLOGY

*Ron Mussieux – Provincial Museum of Alberta*

Tucked away in the badlands near Drumheller, on North Dinosaur Trail, is the Royal Tyrrell Museum of Palaeontology, one of the world's largest paleontological museums. The museum covers 11,200 square metres and, with its "celebration of life" theme, it is one of Alberta's major tourist destinations.

After Joseph B. Tyrrell stumbled across a huge *Albertosaurus* skull in the badlands along the Red Deer River in 1884, Drumheller became a mecca for paleontologists around the globe. For the first century of digging, most specimens were sent to museums outside Alberta for display and research, but with the opening of the Tyrrell Museum in 1985, most of Alberta's dinosaurs remain here.

The museum offers a world class exhibit hall with over 35 complete dinosaur skeletons — the world's largest such display — plus fossils of prehistoric mammals, flying and marine reptiles, and marine invertebrates. A subtropical arboretum houses over 100 plant species, many of which thrived in the dinosaur era. The various galleries explore the Ice Age, the theory of evolution, and man's appearance of earth. The Tyrrell Museum is also a major research centre, and a viewing window into the main laboratory allows us to watch technicians at work. Hands-on displays and computer terminals for visitors make an ancient story modern and interactive.

Interested in dinosaur hunting? The Royal Tyrrell Museum of Palaeontology offers a glimpse into the past, both inside and out. Trails leading away from the Museum take you into the rich bad-

*Ron Mussieux – Provincial Museum of Alberta*

*Modern exhibit is a mixture of mounted skeletons and open dioramas.*

lands environment. There is also a field station affiliated with the Tyrrell Museum at Dinosaur Provincial Park. For information about both these facilities, contact the Royal Tyrrell Museum of Palaeontology.

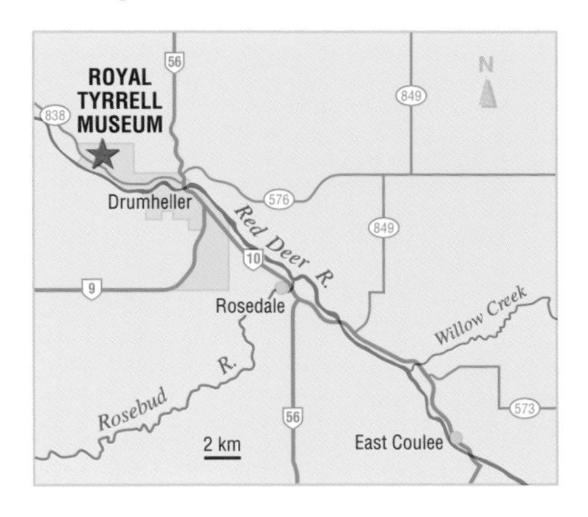

# DRUMHELLER: Minerals Among the Fossils

*Ron Mussieux – Provincial Museum of Alberta*

*Crystals of yellow calcite and colorless quartz fill in fractures in ironstone.*

## HIGHLIGHTS

Although the Drumheller region is world famous for its dinosaur fossils, few people know of the interesting mineral crystals that are also found here. In outcrops along the Red Deer River valley, crystals can be found in the porous portions of dinosaur bones, in fractures in petrified wood, and in the cracks of ironstone concretions. The most common of these minerals are calcite, quartz, and a rare kind of quartz called "pseudo-cubic quartz."

## THE STORY

The Drumheller area is one of the world's most productive dinosaur collecting sites. These fossils are found in rocks of the Edmonton Group that was deposited as an ancient river delta between 75 and 65 million years ago. Sediments continued to be deposited in the Drumheller area, eventually reaching a thickness of 1.3 kilometres above the rocks currently exposed in the valley. Under this thick wedge of sediments the temperature in the sandstone rose to between 38° and 48°C, hot enough to cause minerals to crystallize from the groundwater trickling through the sediments.

The minerals which are found around Drumheller include calcite and various forms of quartz, such as chalcedony, simple and complex quartz crystals, and pseudo-cubic quartz crystals. The calcite often forms small yellow crystals but you may find them as large as two centimetres across; the yellow color is probably the result of small impurities of iron. The fine-grained variety of quartz, called chalcedony, often looks like congealed wax. Milky and clear chalcedony are common in ironstone

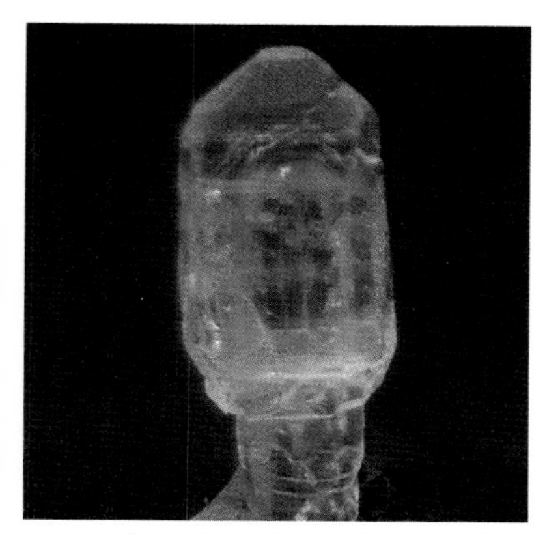

Gregory Baker – Provincial Museum of Alberta
*Quartz "sceptre" crystal.*

nodules but are often coated with coarse quartz crystals or calcite.

Quartz crystals may be found in three basic shapes. The most common is the simple transparent six-sided crystal, which is often found in radiating groups of sprays. "Sceptre" quartz crystals are also quite common. These consist of long crystals with fatter overgrowths of quartz on top, giving the whole crystal the appearance of a regal sceptre.

The rarest quartz found in the area is the pseudo-cubic quartz. It looks like small white cubes that measure up to 10 millimetres across. They are usually found inside ironstone concretions or on petrified wood. This type of quartz is rare because quartz usually forms long, six-sided crystals, not cubes. However, under the pressure and temperatures experienced by the buried sandstone, cristobalite, a mineral chemically identical to quartz, became more stable than quartz. The cristobalite grew in little cubes, but once it was exposed at the surface, it became unstable and eventually changed into quartz. The quartz retained the cube shapes and it is called "pseudo-cubic quartz."

Gregory Baker – Provincial Museum of Alberta
*Spray of radiating quartz crystals.*

Gregory Baker – Provincial Museum of Alberta
*Layered, waxy-like chalcedony topped with quartz crystals on petrified wood.*

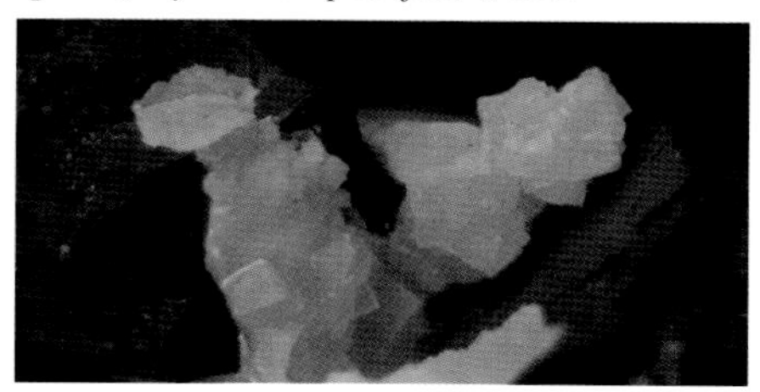

Gregory Baker – Provincial Museum of Alberta
*Intergrown group of pseudo-cubic quartz crystals.*

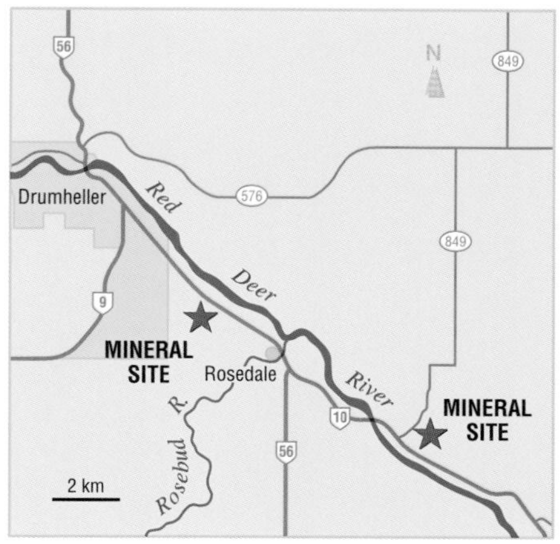

# ROSEDALE BURNT SHALE: The Fires Within

*Ron Mussieux – Provincial Museum of Alberta*

*Quarry of burnt shale near Rosedale. Notice the abandoned coal mining equipment at the top of the hill.*

## HIGHLIGHTS

Looking north across the Red Deer River from Rosedale you will see bright red-orange hillsides. They consist of burnt shale that was baked by burning coal waste heaps, remains of the demolished Star Coal Mine. Baked shale, called clinker, can be found throughout Alberta, and where it is plentiful, is quarried and crushed for use as an ornamental stone. **Do not approach steaming mounds of burnt shale as some are still smouldering and burning beneath a cold, but thin black crust.**

## THE STORY

When coal seams or waste coal piles are exposed to the air, gases are released and the coal may occasionally either spontaneously burst into flames or be ignited by lightning or prairie fires. Once the heat reaches about 1000°C, the surrounding rocks begin to bake into clinker with various shades of red, orange, purple, black, and green. At temperatures over 1117°C, clinker will actually begin to melt and fuse to form what is called paralava. Paralava is often mistaken for volcanic lava, and indeed, it does often contain some high-temperature igneous minerals. Sometimes the paralava will even flow downhill in a fashion similar to a volcanic lava flow. The degree of baking or thermal alteration within a single outcrop is variable, and a hillside may range from slightly baked "clinker" to entirely fused and porcelain-like paralava. Most clinker samples from coal spoil heaps are only melted in part, and some of the red shale may even show the remains of its original bedding.

Gregory Baker – Provincial Museum of Alberta

*The shale has been melted into a glassy clinker, or paralava, that is often mistaken for volcanic lava.*

As long as the methane gas and oxygen are available, coal will continue to burn. Some of the coal seams in the Drumheller valley have been burning since 1911 when the coal mines first opened. Once ignited, a burning coal "front" will advance along the coal seam, heating the nearby rocks as it goes. Most of the burnt shale in the Drumheller region comes from mine waste. Fine coal that was too small to be trapped on the screens was dumped down hillsides where it then mixed with other rock waste that was also discarded from the mining operation. The small coal fragments, with a large surface area in contact with oxygen, have a greater potential for spontaneous combustion.

Burnt shale is often used as an ornamental stone for flower beds, and is occasionally used as road gravel. It is mined by front-end loaders and then crushed to the desired size. Some of the shale is ground quite fine and is used as red clay to mark base paths on baseball diamonds.

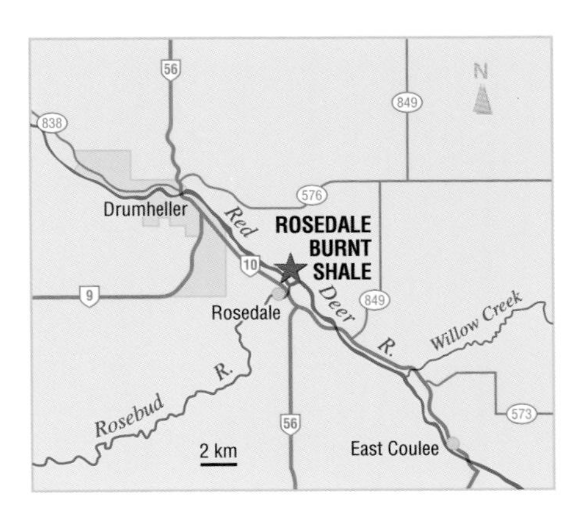

# WILLOW CREEK'S HOODOOS, FOSSIL OYSTERS, AND PETRIFIED TREE STUMPS

*Ron Mussieux – Provincial Museum of Alberta*

## HIGHLIGHTS

Hoodoos are strangely shaped pillars of rock that are produced through erosion by water, wind, and frost. They are common in Alberta's badlands, and those at the Hoodoo Recreational Area are large, accessible, and famous enough to represent Alberta on a special series of twenty-five cent coins. Across Willow Creek from these hoodoos are fossil oyster beds and petrified tree stumps that reflect the high and low water levels of the Bearpaw Sea, the last ancient sea that once submerged much of Alberta.

## THE STORY

Hoodoos form where there is a hard "caprock" which shelters the softer rocks beneath it from erosion. As the soft rocks surrounding the protected sediments erode away, a free-standing pillar is formed. Gradually, however, rain will undermine the caprock which topples over and exposes the softer sediments beneath. Without its protective cap, the hoodoo pillar will rapidly disappear. Hoodoos evolve as fast as one centimetre per year on some faces, and are very fragile.

Hoodoos at this site show three layers — base, pillar, and caprock — that were all deposited during the Upper Cretaceous Period, between 75 and 70 million years ago. The base is red-brown marine shale of the Bearpaw Formation, which was laid down in the inland Bearpaw Sea. The pillar and caprock are sand and clay of the Horseshoe Canyon Formation. They were deposited along the shoreline in deltas and tidal-flats and by rivers that

Ron Mussieux – Provincial Museum of Alberta

*Fossil tree stump weathers out of a coal seam along Willow Creek.*

flowed across this area as the sea shallowed. The caprock contains nearly 40 per cent calcite cement and is therefore more resistant to erosion than the pillar. This hoodoo group consists of 8 to 10 columns ranging in height from one to three metres.

Across Willow Creek from the hoodoos is another section of the Horseshoe Canyon Formation. A scramble across the creek and a 100-metre climb up the slope will bring you to these rocks. Exposed here is a two-metre thick bed of fossilized oyster shells. These shells were broken by storms and washed onto the shore as the Bearpaw Sea deepened briefly and spilled over the delta and tidal-flats. Going farther up the slope, you will see that the oysters are covered by layers of sandstone, shale, and coal. These represent another lowering of the sea level and the resulting influx of river sediments and decaying swamp vegetation that slowly engulfed and buried the oyster shell fragments. The shale and coal contain large, petrified tree stumps, some still in the same position that they grew, some 70 million years ago. Many geologists believe that these trees were drowned by a quick rise in the sea and were preserved in place.

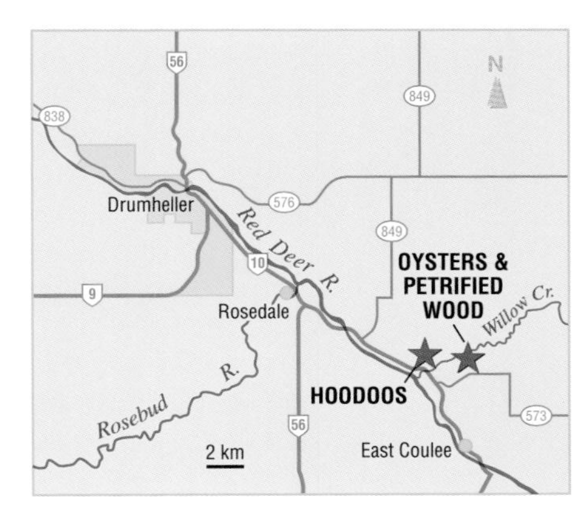

## ATLAS COAL MINE: From Plants to Coal

*Ron Mussieux – Provincial Museum of Alberta*

*The last standing wooden tipple in Canada at the Atlas Coal Mine, East Coulee.*

### HIGHLIGHTS

Even though coal mining has taken a backseat to the petroleum industry in Alberta, it was once a major source of early prosperity in the province. It dramatically affected settlement as towns sprang up, far from major centres, to supply mine sites. In the Drumheller area of the Red Deer River valley, 139 coal mines opened and closed over the years! The Atlas Coal Mine, near East Coulee, was the last of the valley mines to close its doors and is now a Provincial Historic Site.

### THE STORY

Coal is primarily carbon mixed with some hydrogen, oxygen, and nitrogen. Coal-bearing formations underlie 300,000 square kilometres of Alberta. Where did all this coal come from? During the Cretaceous Period, parts of Alberta were steamy, subtropical swamp, similar to the Florida everglades today. This was the "era of the dinosaurs." In the stagnant waters, layers of decomposing vegetation accumulated to form thick peat mats. Over time, the peat was buried under heavy deposits of sand and mud and was slowly compressed and heated into coal. About 10 metres of compacted plants would become one metre of coal. The longer it was compressed, the more moisture and gases were driven out,

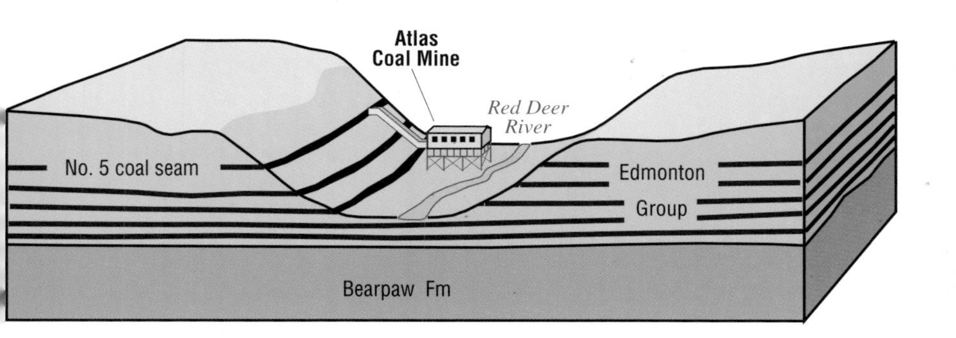


leaving only carbon behind. As the carbon content increased, the heat value or rank of the coal increased.

Alberta's coal is geologically young and therefore we might expect its rank to be quite low. The great pressures applied to the coal during the building of the Rocky Mountains, however, increased its rank, particularly in the mountains and foothills. Thus, as we go west in Alberta towards the mountains, the rank of coal increases.

Unfortunately though, the higher rank coals of the mountains and eastern foothills are relatively expensive to mine because the seams are often steeply dipping, folded and faulted. Blasting and drilling are required and a lot of the coal breaks into coal dust. The coal of the plains, however, underwent little disruption during mountain building, and the seams are flat-lying and covered only with soft sediments. Today, the plains coal is mined to produce electricity at the generating stations near Wabamun, Forestburg, and Sheerness.

In the past, many of the seams in the Drumheller area were mined underground and you can see several of these seams along the river valley. The Atlas Coal Mine opened in 1928 and for 51 years mined low rank sub-bituminous coal to heat homes and offices in the Drumheller area. The mine site now provides information on equipment, mining techniques, and the last wooden coal-sorting tipple in Canada.

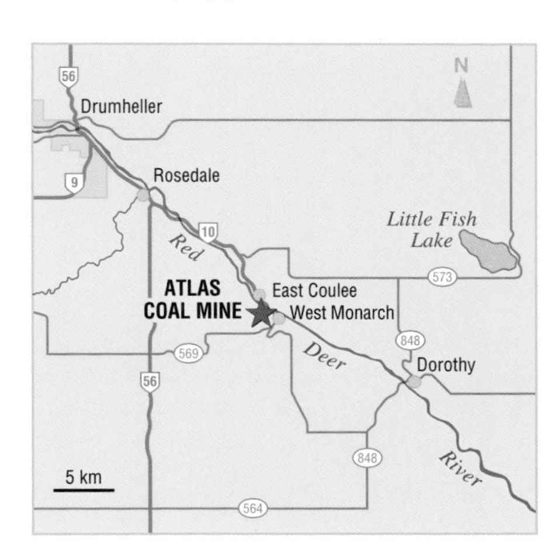

# BENTONITE, OR "POPCORN ROCK," FROM DOROTHY

Ron Mussieux – Provincial Museum of Alberta

*Red Deer River valley at the Dorothy Bridge. The 10-metre thick white bed exposed near the valley floor is composed of impure bentonite.*

## HIGHLIGHTS

The crumbly, grey, popcorn-like rock that litters the badlands is a type of clay called bentonite. When it becomes wet, it soaks up water like a sponge and turns into a slimy mud that often sends the unsuspecting hiker sliding downhill. It is best exposed near Dorothy, where a thick bed can be traced for several kilometres along the banks of the Red Deer River. Bentonite is so widespread in Alberta that beds of bentonite are a useful tool for correlating and dating rock layers.

## THE STORY

Bentonite is composed mainly of the clay mineral montmorillonite, which is formed by the chemical alteration of volcanic ash. Around seventy million years ago, plate collisions produced numerous active, erupting volcanoes, likely in the Yellowstone area and southern British Columbia. These volcanoes were very explosive, erupting like Mt. St. Helens in 1980, blowing ash over huge areas. The airborne ash floated eastwards and settled over much of western Canada. It landed in the salty Cretaceous sea, on the marshy shores, and on dry land. The volcanic ash that landed in the seawater was more likely to alter to thick beds of bentonite than the ash that landed on the land.

Montmorillonite absorbs water and can swell up to 10 times its dry volume becoming a slippery, greasy clay. Wet bentonite below the ground surface acts as a lubricant promoting slumping along river valleys. As a result, it plays an important role in the rapid erosion of the badlands and the formation of their unique topography. Bentonite also influenced the coal mining industry in the Drumheller area. In the Atlas Coal Mine, for example, a bentonite bed formed the roof of the mine shaft creating a dangerous situation for miners when it became moist.

Because of its absorptive properties, bentonite is used in the manufacturing of many products. Historically, this clay was used by Aboriginal Peoples and at various Hudson's Bay Company posts as soap because it works into a lather, earning it the name "soap clay." Today, its most important use in Alberta is as drilling mud, where its special swelling properties make it ideal for sealing porous rocks.

The 10-metre thick bed of bentonite near Dorothy is the thickest and most extensive outcrop in Alberta. Although there is a substantial amount of bentonite in this bed, impurities make it low-grade and it has not been mined. Bentonite mining in Alberta has been intermittent over the years but today all bentonite is imported.

*Ron Mussieux – Provincial Museum of Alberta*

*Bentonite shrinks upon drying to form a popcorn-like surface.*

*Royal Tyrrell Museum of Palaeontology*

*Tyrrell Museum employee Darren Tanke takes the big slide on a greasy bentonite slope.*

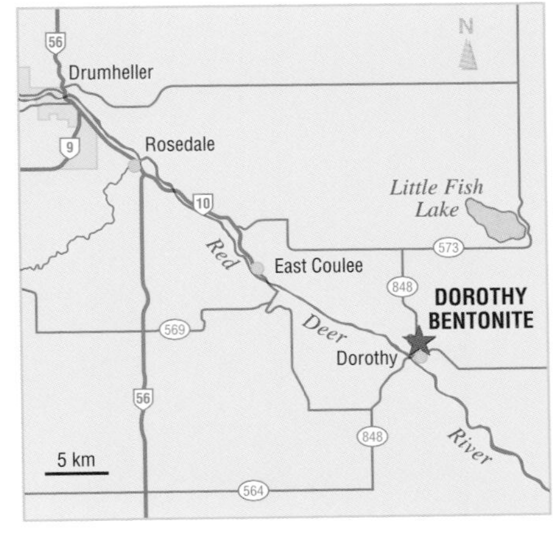

# DINOSAUR PROVINCIAL PARK BADLANDS:
## A World Heritage Site

*Ron Mussieux – Provincial Museum of Alberta*

*Sunset in the badlands adds a red color to these ancient river sands.*

## HIGHLIGHTS

Badlands are a landscape of deeply eroded rocks, with little vegetation, that is a wilderness of slopes, scarps, and narrow, twisting valleys. Although badlands are scattered throughout Canada, they are most rugged and spectacular in Dinosaur Provincial Park. This park is best known, however, as one of the most important dinosaur fossil localities in the world. In 1979, Dinosaur Provincial Park earned the status of a UNESCO World Heritage Site because of its abundance and diversity of fossils, its river habitat, and its badlands. With this designation, it joined the ranks of other globally significant areas such as the Great Barrier Reef and the Galapagos Islands.

## THE STORY

### THE ANCIENT ENVIRONMENT

Between 100 and 70 million years ago, the interior of North America was covered by a shallow, subtropical sea that, at times, extended from the Arctic Ocean to the Gulf of Mexico. Dinosaur Park lies close to the former shoreline and was an environment of moist deltas with lush swamps. The rock layers now exposed along the Red Deer River valley in Dinosaur Provincial Park tell the story of the ancient sea's fluctuating water levels and the changing positions of its shoreline. Geologists have divided these rock layers into three units, or formations, each representing different ages and depositional environments.

About 30 metres of the lowest, there-

'ore oldest, exposed unit is part of the Olman Formation. It consists of pale yellow sandstone, siltstone, and rust-colored ironstone deposited about 77 million years ago. At that time, the sea was 400 kilometres to the east, and Dinosaur Provincial Park was a low-lying flood plain across which rivers and streams wove their way eastward to the sea. The mountains rising to the west supplied huge quantities of sediment to the rivers which deposited it over eastern Alberta and Saskatchewan. The climate was hot and dry and frequent floods deposited sediments and broken dinosaur bones over the area. Complete dinosaur skeletons are rarely found in this formation.

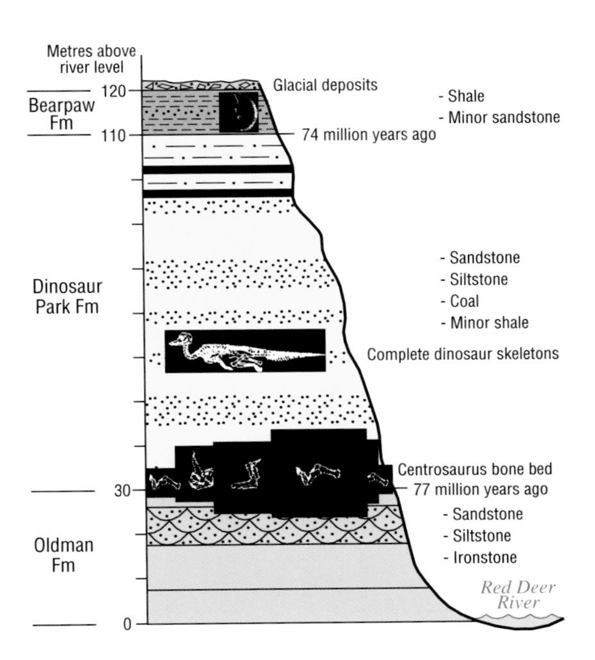


About half a million years later, the environment became humid and warm as the sea level rose and the shoreline of the inland sea advanced westward towards Alberta. The rocks that represent this period are about 80 metres thick and consist of red sandstones and siltstone, some claystone, and thin layers of coal. These sediments were deposited by meandering rivers in the lush swamps, deltas, and estuaries. Dinosaurs flourished in this subtropical climate. The complete fossil skeletons collected in the park come from these rocks, which are aptly called the Dinosaur Park Formation.

Two million years later, the sea made its final advance over south and central Alberta depositing the dark brown and black marine sediments of the Bearpaw Formation. Here, this formation is 10 metres thick and contains abundant marine fossils, such as ammonites, clams, shark teeth, and crayfish.

About two million years ago, the climate cooled and Alberta was in the grip of the Ice Age. The badlands in Dinosaur Provincial Park began to form around 15,000 years ago as the glaciers finally melted from this area. The scouring action of the glaciers exposed the soft, easily eroded rocks we see today along the river valley. Torrents of glacial meltwater cut them deeply, creating steep-sided channels, one of which is now occupied by the Red Deer River. This set the stage for the development of the present-day badlands.

## EROSION AND THE LANDSCAPE TODAY

The eerie landforms in the badlands today are the result of extremely rapid erosion, no longer by water from melting glaciers, but by rain and running water. There are several factors that combine to make the badlands in Alberta so vulnerable to erosion. First, the rock layers are soft, weak and easily eroded. Second, the badland sediments contain numerous layers of bentonite clay which, when wet, swells rapidly. During a rainstorm, swollen blocks of this clay will quickly slide down the hillsides. Third, occasional, heavy rainstorms in an otherwise semiarid climate allow for quick erosion by surface water. These three factors, in addition to steep valley walls ensure rapid erosion. The average rate of erosion in this Park is about 0.5 centimetres per year, which is 1000 times greater than the rate in the Rocky Mountains! This is incredibly fast by geological standards making badlands perfect places to study erosional processes.

Dinosaur Provincial Park contains numerous landforms produced by erosion. The most familiar ones are hoodoos. These may take on shapes of pedestals or toadstools, and are produced in areas where a layer of soft rock is overlain by a harder rock. As the surrounding rocks are eroded away, a pillar of soft rock with its protective caprock is left behind. Rill erosion, found on steep sandstone hillsides, looks like a branching network of small grooves. They are formed after a rainstorm as the water washes away the soft sediments on the hillside. With each rainfall, the rills continue to extend and deepen until they form pipes and tunnels. Pipes are vertical shafts that channel the water down into the hills and connect into subsurface tunnels. Through these pipes and tunnels, the water will eventually reemerge at the base of the hill. When tunnels collapse, they form either sinkholes on the surface, or new gullies. As the badlands continue to be extended into the prairies, mostly by gully erosion, other dramatic landforms such as pinnacles, buttes, and coulees are formed. This rapid erosion in the badlands of Dinosaur Provincial Park is instrumental in revealing the skeletons of the long-dead

*Rills form on hard sandstones.*

*Erosion of rocks of varying hardness produced the "camel."*

*Rain falling on harder sandstone will produce teepee-shaped buttes.*

*Underground drainage system cut by water in a process called "piping."*

*Heavy rainfall fills the usually dry stream beds.*

*Ron Mussieux – Provincial Museum of Alberta*

## DINOSAURS

So far, over 30 different species of dinosaurs, representing six dinosaur families have been found, as well as over 80 species of other vertebrates, such as crocodiles, lizards, frogs, mammals, flying reptiles and birds. The lush, subtropical climate was perfect for dinosaurs, and harsh winters were still unknown. In this climate giant redwoods, sable palms, *Gingko*, and large ferns flourished. The most common dinosaur group in the park was the duck-billed dinosaurs, or hadrosaurs. Thousands of them, such as *Edmontosaurus* and *Saurolophus* munched on the leaves of cypress trees and oaks in the swamps and estuaries. Other plant eaters, such as the small *Stegoceras*, the club-tailed ankylosaurs, and *Centrosaurus* were plentiful. Of particular interest in this park are spectacular deposits of *Centrosaurus* bone beds, where thousands of individuals are spread across eight square kilometres. Some paleontologists believe they likely drowned in a flood and their carcasses were swept downstream and piled together. Herds of other ceratopsians, or rhinoceros-like dinosaurs, also roamed through the park. They had shearing teeth and were able to browse on the toughest-leaved plants.

These plant eaters fell prey to a variety of predatory, carnivorous dinosaurs. One predatory dinosaur, the bipedal, bird-like *Dromaeosaurus*, had a large brain, was very agile and fast, and with the long claw on its hind inner toe, was likely able to disembowel its prey. *Albertosaurus*, a smaller relative of *Tyrannosaurus*, also frequented this area and thrived in this environment.

The parking lots near the Park gate are excellent viewpoints over the vast expanse of badlands. The best place to begin your exploration of the park is at the Field Station of the Royal Tyrrell Museum of Palaeontology. From here, there are guided bus and hiking tours into the restricted areas. There are also two short interpretive trails that you can explore on your own — the Badlands Trail starts just east of the campground, and the Cottonwood Flats Trail starts from behind the campground. Dinosaur Park is a unique locality to get a glimpse

*Ron Mussieux – Provincial Museum of Alberta*

*Rare dinosaur skin impressions in ironstone. The gut cavity is filled with wood fragments.*

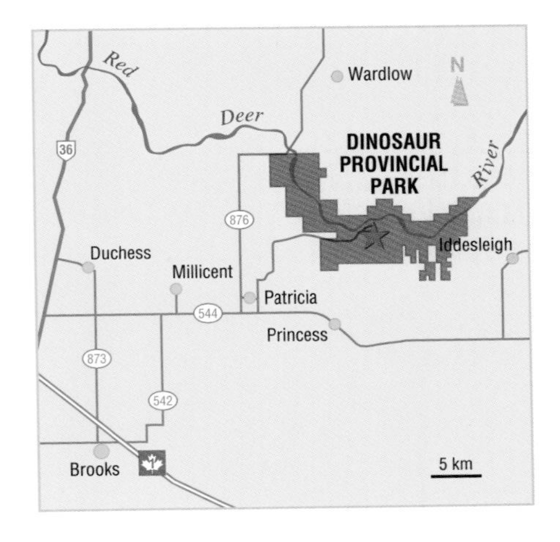

# MEDALTA POTTERY: From Clay to Stoneware

*Ron Mussieux – Provincial Museum of Alberta*

*Metal-strapped beehive kiln at the Clay Products Interpretive Centre, which houses "The Great Wall of China." These kilns and those at the nearby Medalta Historic Site are rare and are being preserved.*

## HIGHLIGHTS

Medicine Hat has historically been the major centre of Alberta's clay and ceramic industry because it is close to extensive clay deposits and abundant, cheap natural gas. In fact, the city sits on a gas field, which led writer Rudyard Kipling to describe it as "a city with all hell for a basement!" Medalta Potteries (1916-1954) was the first major factory of the ceramics industry in Canada, and in recognition of its historical and cultural contribution, its kilns and buildings were declared a National and Provincial Historic Site in 1975. The Medalta Site provides tours, displays, and exhibits old machinery of the ceramics industry.

## THE STORY

Clay products are among humanity's first manufactured objects and even today they are invaluable to modern society. Clay ranks as one of the leading industrial minerals, both in tonnage and in value. Why? Clay has the ability to hold a shape because of its unique atomic structure. It is a fine-grained, earthy material that is composed of tiny plate-like minerals. It can be moulded into any shape when wet, and upon firing it becomes a permanent hard mass that can be harder than metals. Different clays have different properties and are therefore not used for the same purposes.

Even though clay is plentiful in Alberta, most is of poor quality and only suitable for making brick and tile. The problem with Alberta clay is that it contains bentonite, which shrinks, cracks, and warps as it dries. For this reason, Medalta brought in its high quality clay from nearby quarries in Eastend and Willow, Saskatchewan. Today, the clay works in Medicine Hat still receive most of their clay from there, plus some from quarries just north of Cypress Hills.

Medalta used three main steps to turn clay into a finished ceramic product: shaping, drying, and firing. Firing was done in gas-fed kilns, which are similar to ovens, at temperatures between 800° and 2000°C. Four bee-hive kilns that were built in 1920 remain at Medalta; kilns from this period are very rare. Many finished ceramic objects were coated with a glaze for decoration and protection. Interestingly, between 1937 and 1940, Medicine Hat potteries used a radioactive uranium-oxide glaze for a fabulous deep orange color. This was discontinued with the onset of the Second World War.

Medalta was a household word in the 1920s and '30s and over the years the company made a variety of items including stoneware, crocks, vases, porcelain, hotel china, lamp bases, and bricks. For the first half of this century, it supplied over half of Canada's pottery and was the first western Canadian company to ship manufactured goods to eastern Canada. It was the major supplier for CPR hotels and train dining cars, and during the war sent cups and dishes to our Canadian Armed Forces. In 1954, Medalta Potteries closed, and its ceramic pieces are now collectors' items.

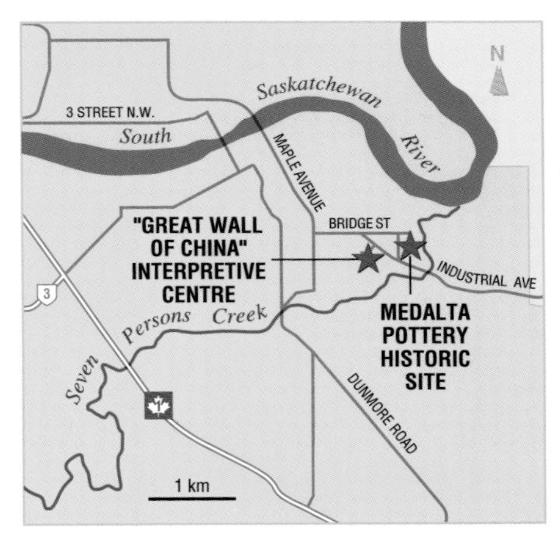

*Gregory Baker – Provincial Museum of Alberta*

*Crocks and jugs were major products of Medalta Potteries.*

# CYPRESS HILLS: An Oasis in the Prairies

*Ron Mussieux – Provincial Museum of Alberta*
*Reesor Lake and the tree-clad slopes of Cypress Hills (above).*
*Ron Mussieux – Provincial Museum of Alberta*
*The upper slopes of the Cypress Hills consist of a well-cemented gravel.*

## HIGHLIGHTS

In 1859, John Palliser of the Palliser Expedition wrote in his journal that Cypress Hills was "an oasis in a desert, an island in a sea of grass." The forested hills rise over 800 metres above the surrounding plains and are the highest point between the Canadian Rockies and Labrador. Cypress Hills has been an upland area for millions of years mainly because it is capped with a tough, erosion-resistant layer of conglomerate.

## THE STORY

The conglomerate "caprock" of Cypress Hills, called the Cypress Hills Formation, is composed of pebbles and cobbles that were eroded from the uplifted Sweetgrass Hills, Bearpaw Mountains, and Highwood Mountains of Montana, and carried north in broad, turbulent rivers. The rivers meandered across the plains of Alberta and Saskatchewan and deposited sheets of gravel which protected the sediments beneath from erosion. Vast amounts of sediments surrounding the gravel deposits were eroded away until gravel-capped plateaus, such as Cypress Hills, were left standing above the plains. In other words, the top of the Hills was once the bottom of an ancient river bed. Spectacular deposits of this conglomerate can be seen at the western end of Cypress Hills where it is over 100 metres thick! This conglomerate is famous for its wealth of vertebrate fossils, which indicate it is between 44 and 35 million years old. Due to its elevation, Cypress Hills became the drainage divide that separated the Saskatchewan and Milk (Missouri) River systems.

Cypress Hills was only partially covered by ice during the last Ice Age. When the massive, continental glacier flowed southward towards the Hills, it divided and flowed around them, leaving the top 100 metres as an island of land towering above the surrounding ice. This unglaciated area is significant because it is one of the oldest surfaces unaltered by ice in Western Canada, basically unchanged since it formed millions of years ago. As the glaciers melted, huge volumes of water carved out deep channels and coulees along the north and west slopes. Today, Elkwater Lake, Reesor Lake, Battle Creek, and Spruce Coulee lie in these channels.

Cypress Hills, with greater precipitation and lower temperatures than the prairies immediately around it, features a diverse ecosystem more typical of the Rocky Mountains. Rainwater quickly soaks into the porous conglomerate and flows underground until it eventually trickles out along the hillsides as springs. Lush forests and vegetation flourish here, even during the dry, prairie summers. Often the soil becomes oversaturated with water, and slumping of the hillsides is common. Thus, while the conglomerate helps maintain the elevation of the Hills, its porosity is a major contributing factor to landslides on its slopes.

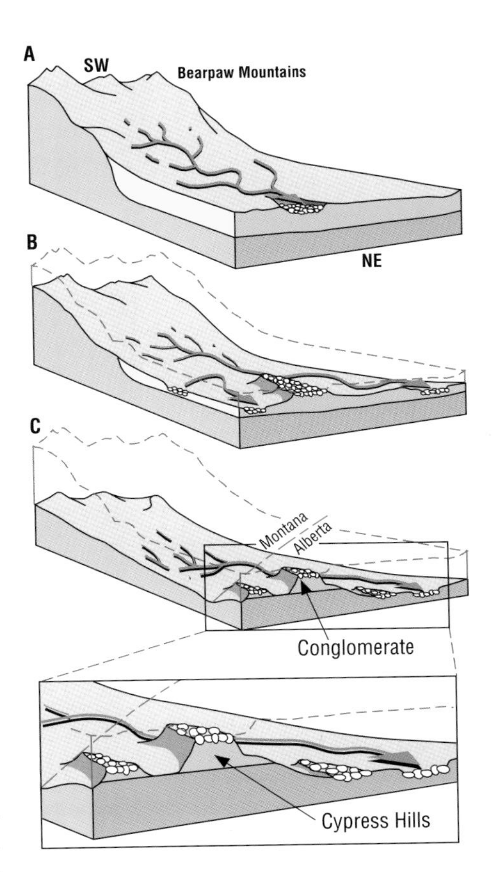

# THE STRANGE SPHERES OF RED ROCK COULEE

*Natural Resources Service*

## HIGHLIGHTS

The eerie moon-like landscape of Red Rock Coulee is created by badlands, hoodoos, and numerous huge, rust-colored boulders, called concretions. Many of these concretions are 2.5 metres across and are by far the largest, and best developed, in the province. Looking across the coulee, you can see concretions in many different stages of emergence as the softer shale of the Bearpaw Formation is slowly eroded away exposing the concretions. Some are still partly buried with only their tops visible, while others are completely exposed and scattered along the coulee walls.

## THE STORY

Concretions are a type of sedimentary structure that arouses a great deal of interest because of their widespread occurrence and variety of unusual shapes. They can show such remarkable symmetry that they have been erroneously mistaken for dinosaur eggs and brains, meteorites, and even fossil animals!

A concretion often begins as a dead organism, such as a leaf, bone or shell, that is buried in a soft "host" sediment such as sand or clay. The more homogeneous, or pure, the host sediment, the more favorable the conditions for large, spherical concretions. As subsurface groundwater trickles through the sediment, dissolved minerals are attracted to the fossil because it is chemically different from the host rock. The minerals accumulate and precipitate around the fossil, cementing together sediment grains until a dense rock structure, or concretion, is formed. Concretions are harder than the surrounding sediments and therefore resist erosion. We know that iron oxide was one of the cementing minerals at Red Rock Coulee because of the striking red color of the concretions, which contrasts with the enclosing grey shale.

Once precipitation of minerals around the nucleus has begun, the concretion continues to grow outwards, producing a definite concentric layering, similar to an onion. This concentric structure is a key identifying characteristic of a concretion. Occasionally, a well preserved fossil can still be found in the centre of a broken concretion, although some concretions will form around an inorganic nucleus.

Concretions range in diameter from a few millimetres to several metres. We can only speculate why they were able to grow so large at Red Rock Coulee. There were likely long periods of uninterrupted growth with mineral-rich groundwater flowing continuously through the shale beds. In a few cases, where a concretion is only partly exposed, it is possible to see the shale beds are bent upwards over it and downward under it. This tells us that these concretions formed considerably after the Bearpaw shales were deposited. Red Rock Coulee has been designated as a Natural Area because of its beauty and fascinating rock forms.

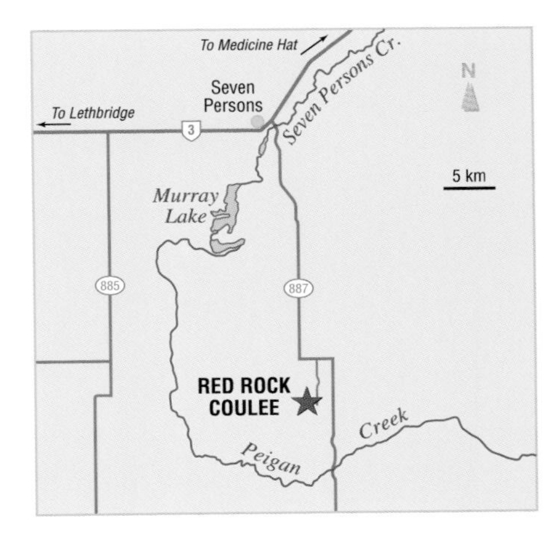


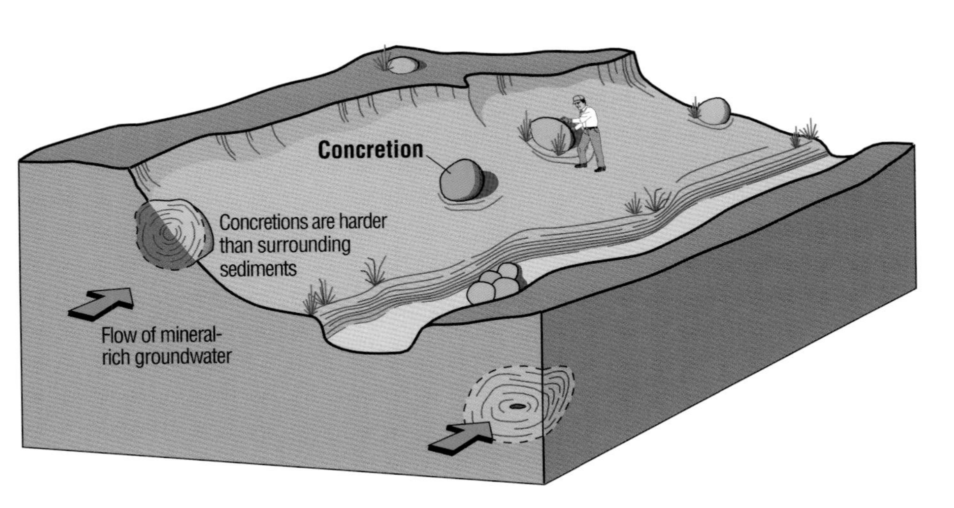


*Hard concretions weather out of the soft shale of the Bearpaw Formation.*

# MILK RIVER DYKES: Walls of Once-molten Rock

*Don Taylor – Provincial Museum of Alberta*

*Cockscomb Dyke, the most easterly of the dykes, is in the Milk River Natural Area.*

## HIGHLIGHTS

Southeast of Writing-On-Stone Provincial Park is a group of rare rocks, some of which have the appearance of ancient, decaying stone walls. These are the Milk River Dykes, eight masses of igneous rock that stand out in the river valley and coulees because they are harder, and more resistant to erosion, than the surrounding sandstones. Many of the dykes have been given local descriptive names such as Black Butte, the Roman Wall, and the Cockscomb. Three of the more accessible dykes are indicated on the map: Black Butte, which Secondary Highway 500 crosses; Cockscomb Dyke, which is about a 1.5-hour drive from Writing-On-Stone Provincial Park; and the McTaggart Coulee Dyke, which is about 500 metres north of Secondary Highway 500.

## THE STORY

The Milk River Dykes were formed about 50 million years ago when molten rock, called magma, rose through near-vertical fractures in the crust and hardened before reaching the earth's surface. The main igneous rock body is located south of these dykes, and forms three large buttes known collectively as the Sweet Grass Hills of Montana. Despite their conical shape, these hills are not volcanoes but rather hard, igneous rocks that cooled below the earth's surface and have been exposed by the erosion of overlying soft bedrock. The Sweet Grass Hills are surrounded by many dykes, including the Milk River Dykes, all of which are part of the same igneous intrusion.

The composition of the rocks forming the Milk River Dykes varies. The seven easterly dykes located close to the East Butte of the Sweet Grass Hills, are dark-colored, rich in iron and magnesium, and often contain large crystals of dark brown mica. The eighth dyke, the McTaggart Dyke, is a lighter-colored rock that contains the white mineral plagioclase.

Within the dykes are inclusions of other rocks that were broken free from the wall rock as the magma rose up through the crust. These foreign rocks give us a great deal of information about the depths from which the magma rose. Examples of the inclusions are fragments of gneiss from the underlying Shield, and metamorphosed limestones, now marble, that contain fossil sea lily stems. We know this magma came from deep within the earth, perhaps as deep as the upper mantle, because we also find small blocks of olivine-rich rocks in the dykes. Olivine is the most common mineral making up the rocks of the upper mantle. Because the mantle is the ultimate source of diamonds, these dykes were staked and explored for diamonds. So far, only one micro-diamond has been reported.

Many of the dykes are on private ranches and permission must be obtained before exploring them. The Milk River Dyke is in the Milk River Natural Area. If you are planning on collecting rock specimens from here, it is advisable to contact Alberta Environmental Protection Services for permission.

*Don Taylor – Provincial Museum of Alberta*

*This once-molten rock now stands out like a wall in McTaggart Coulee.*

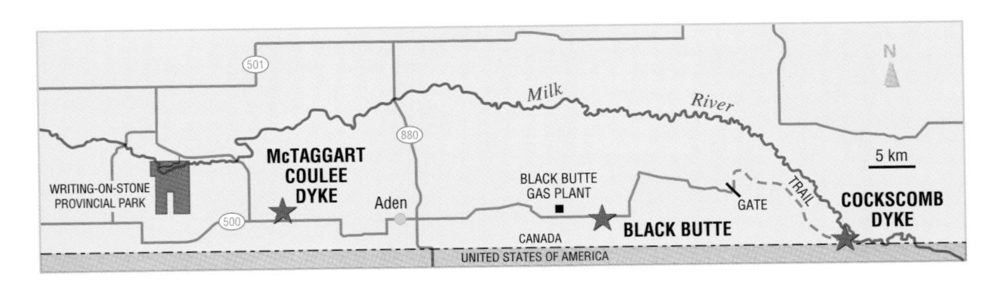

# WRITING-ON-STONE PROVINCIAL PARK

Ron Mussieux – Provincial Museum of Alberta

*Cliff-forming sandstones of the Milk River Formation.*

## HIGHLIGHTS

At Writing-On-Stone Provincial Park, the meandering Milk River has carved deeply into the prairie, exposing the underlying sandstone bedrock. Water and wind sculpt this soft rock into a variety of fantastic shapes, creating a dramatic, surreal landscape of cliffs and massive outcrops. Cliff faces are covered with the largest concentration of Plains Indian "rock art" in North America, and in order to protect it, this park was established in 1957.

## THE STORY

Erosion of the Milk River valley began between 17,000 and 18,000 years ago towards the end of the last Ice Age. A melting glacier north of Cypress Hills had blocked off normal drainage, and the water was then diverted southwards where it carved the present Milk River valley. Huge volumes of water rapidly enlarged the valley, which in some areas of the park is over 400 metres wide and 60 metres deep! Today, the valley of the Milk River is so deep and wide that is seems surprisingly mismatched for its river. This is a classic example of an "underfit stream" — the modern Milk River is simply too small to have eroded the valley we see today.

The cliffs and massive outcrops along the river valley are 80-million-year old Cretaceous sandstone, with thin layers of shale and ironstone. These outcrops are highly fractured and jointed, and are often only weakly cemented. Erosion acts on these weaknesses and erodes the rocks to bizarre castle-like shapes and hoodoos. Hoodoos are pillars of rock that form in badland areas where there are layers of harder, erosion-resistant rocks overlying softer rocks. The harder layer acts as a caprock that protects the underlying layer from erosion. In this area, the caprock is strongly cemented sandstone.

Extensive pitting of the cliff faces, called tafoni, is also a widespread erosional feature here, and is the result of unevenly cemented sandstone. Rain, and to a lesser extent wind, slowly etch out the weaker sandstone surfaces, creating a "honeycomb" structure that can range from a few centimetres to several metres across. Erosion is a powerful and ongoing force in southern Alberta because of the extreme variations in rainfall and temperatures. This natural process is destroying the irreplaceable rock art and research is underway to find practical ways of stabilizing the cliff faces.

The soft sandstone of the steep valley walls is easily carved by side streams forming deep, narrow gorges and coulees. This rugged badland landscape is quite different from that in Dinosaur Provincial Park, where clay-rich rocks are sculpted into round formations. At Writing-On-Stone, the four kilometre (return) Hoodoo Interpretive Trail takes you along the cliffs for an excellent view of the unusual rock outcrops and the mysterious rock paintings and carvings.

*Jack Brink – Archaeological Survey*

*Aboriginal Peoples carved shield figures in the soft sandstone cliffs.*

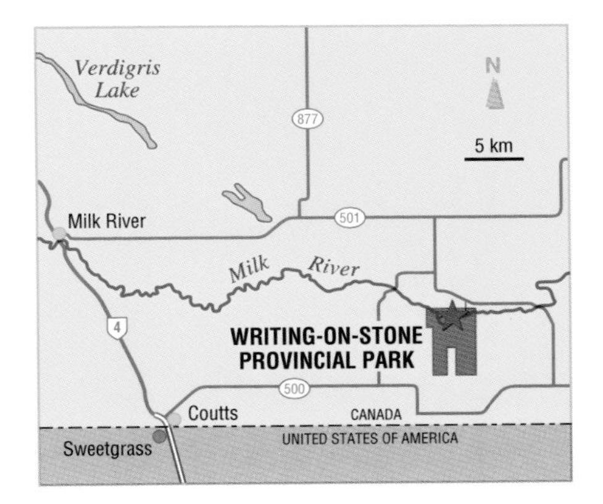

*Ron Mussieux – Provincial Museum of Alberta*

*Additional calcite cement in the thinly bedded sandstone layers makes them harder than the underlying thicker sandstone unit and results in hoodoo development.*

# DEVIL'S COULEE DINOSAUR EGG SITE

*Phil Currie – Royal Tyrrell Museum of Palaeontology*
*Devil's Coulee Egg Site. Notice the excavations midway up the slope (right).*

*Royal Tyrrell Museum of Palaeontology*
*A cluster of dinosaur eggs from Devil's Coulee (left).*

## HIGHLIGHTS

On June 23, 1987, dinosaur nests were discovered in Devil's Coulee near Milk River that contained fossilized eggs with exquisitely preserved bones of unhatched babies. These nests, which also contained bones of hatched and still nestbound babies, were built around 75 million years ago by hadrosaurs, or duck-billed dinosaurs. This exciting fossil discovery is an important key to deciphering the social behavior and physiology of dinosaurs. We now know that some types of dinosaurs actively cared for and protected their young after they hatched.

## THE STORY

Devil's Coulee has yielded many hadrosaur nests from a variety of sites. The nests, found in mudstone, have raised mud rims and are closely spaced, about a hadrosaur length apart, or 12 metres. As well, they are stacked on top of each other with about 0.5 metres of sandy clay between them. The reconstructed eggs are quite large, with dimensions of about 18.5 x 20 centimetres, and have a volume of 3.9 litres. Most of the embryonic material from this location is uncrushed and some fetuses have articulated fossil skeletons.

Paleontologists can learn a remarkable amount from these nests. The nests were built in sediments that indicate that duck-billed dinosaurs lived in mud-covered lowlands and deltas. The proximity of the nests to each other suggests these dinosaurs nested in colonies, as penguins do today, while the stacking of nests tells us the mothers returned to the same spot to nest over a long period of time. Worn teeth in the embryos indicate the teeth were functional before hatch-

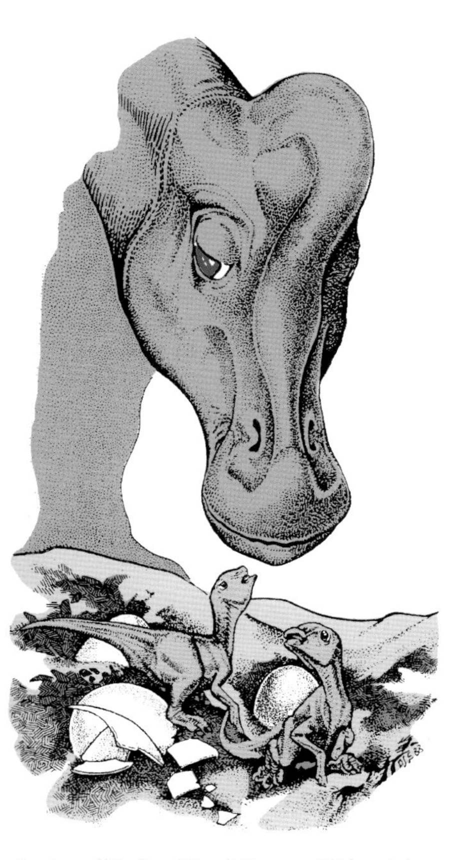

Courtesy of The Royal Tyrrell Museum of Palaeontology.

ing. The large size of the babies still living in nests tells us they grew quickly and therefore may have been warm-blooded. It appears babies were nest-bound until they had at least doubled in size (to well over a metre long), thus the parents must have brought the babies food. The bones also help us understand how these dinosaurs grew from tiny embryos to 10-metre long, four-tonne creatures in only five years.

Beautifully preserved fossils of adult dinosaurs from the Cretaceous Period have been found throughout much of southern Alberta. Well preserved eggs and bones of juveniles are rare however, because the soil chemistry was generally slightly acidic and the delicate shells and bones were dissolved. For some reason, though, the Milk River area was less acidic which helps explain the spectacular preservation of the embryos and eggs. Unfortunately, this different level of tolerance to acidity means that adult and juvenile bones are often not found together and much of the "family life" of dinosaurs remains a mystery.

Tours leave for this site from the Devil's Coulee Interpretive Centre in Warner.

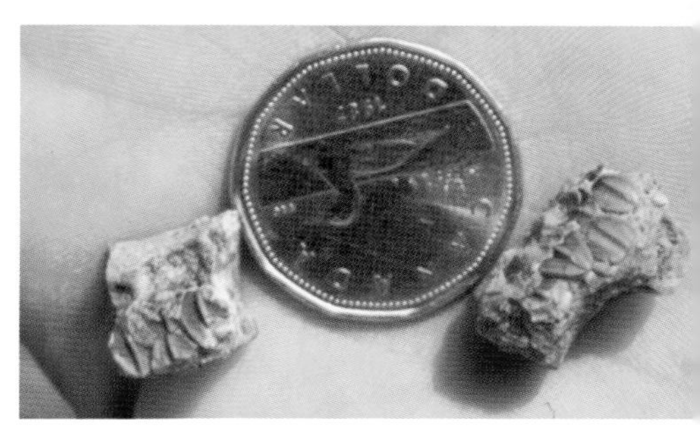

*Royal Tyrrell Museum of Palaeontology*

*A fragment of a baby duck-billed dinosaur jaw with teeth.*

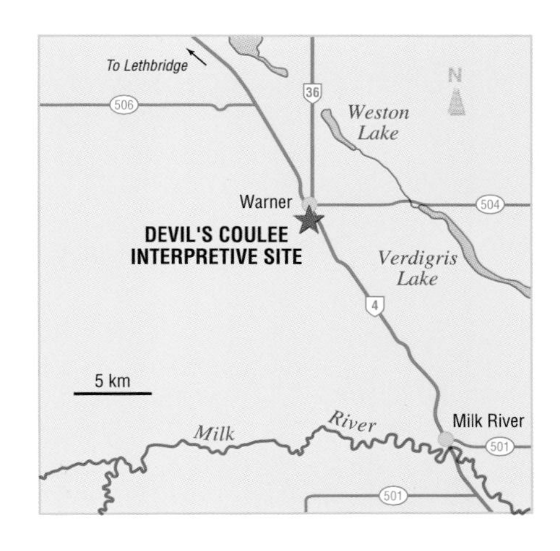

# AMMOLITE: Alberta's Gemstone

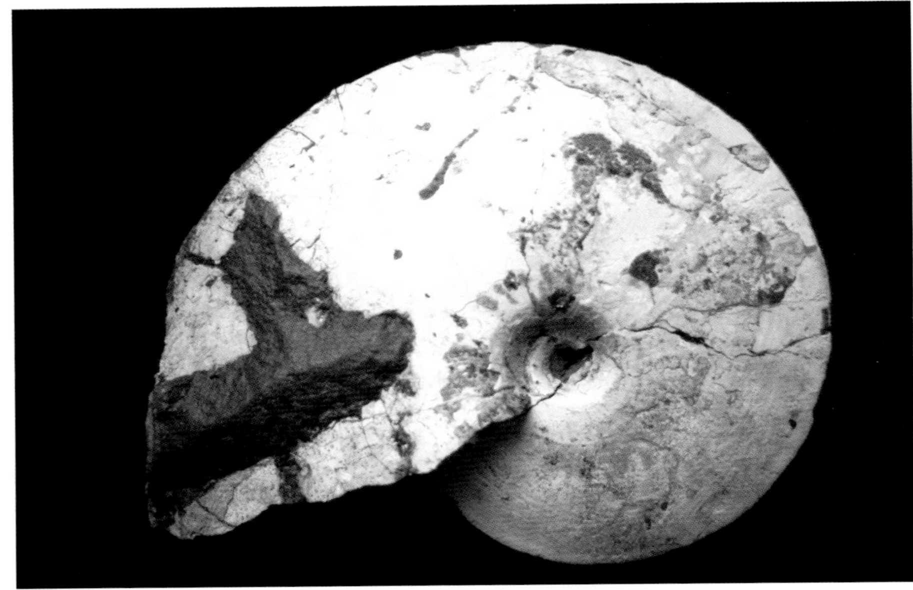

*Gregory Baker – Provincial Museum of Alberta*

*Beautifully preserved coiled ammonite.*

## HIGHLIGHTS

Ammol*ite* is the gemstone name given to the beautiful iridescent fossilized shell of extinct ammo*nites*. Southern Alberta is the only region in the world where shells reaching gem quality have been found. The St. Mary River of southern Alberta is a particularly noteworthy fossil locality because it not only provides the prized ammolite, but also some of the best preserved large ammonites in the world, some reaching one metre in diameter!

## THE STORY

Ammonites are a group of carnivorous molluscs which lived in many oceans worldwide from the Devonian to the Cretaceous periods. Their soft bodies were enclosed in chambered shells which were usually coiled. Because the ammonites attained almost worldwide distribution and had a geologically short life span, they are important guide fossils that accurately date the rocks they are found in.

During the Late Cretaceous Period, southern Alberta was covered by a subtropical inland sea where ammonites flourished. The ammonites that provide the raw material for gems are about 70 million years old, and are found in the marine shales of the Bearpaw Formation, mainly along the St. Mary River, west of the town of Magrath. Geologists are uncertain why commercial quality ammolite is limited to such a small area along this river, but perhaps a combination of rapid burial, heavy pressures, and unique chemical reactions helped preserve the iridescent shell structure. Less than 10 per cent of shells

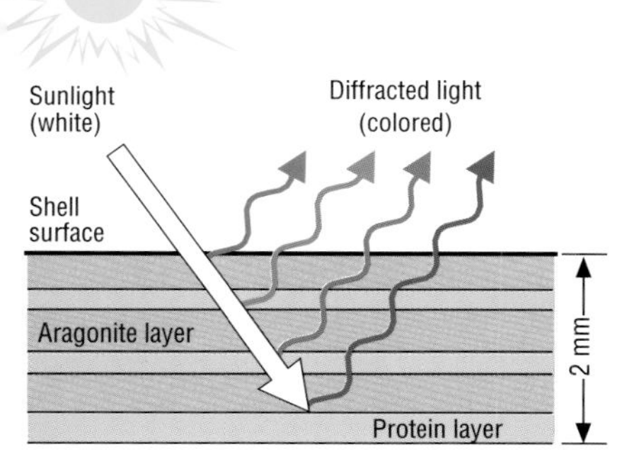


found here are of gem quality, and only 20 per cent of these can be used for jewellery. By Alberta law, only shell fragments can be used for jewellery, complete fossils must be kept intact.

Ammonite shells consist mainly of thin alternating layers of the mineral aragonite and a protein matrix of conchiolin. When light shines on the ammonite shell, it is split by the layers into brilliant reds, greens, yellows, and blues. These colors are then reflected back in essentially the same process whereby a prism splits sunlight into a spectral rainbow. Red and green are the most common colors seen in ammolite. Cracking of the shell over millions of years produces the stained-glass window appearance so characteristic of ammolite.

Ammolite is thin and brittle, so it is usually left on its shale backing for reinforcement. Because it is quite soft, it is also capped with a much harder mineral, such as quartz or spinel, in the production of jewellery.

Please remember that mining ammolite is a commercial operation and some of the land along the St. Mary River is leased for these operations.

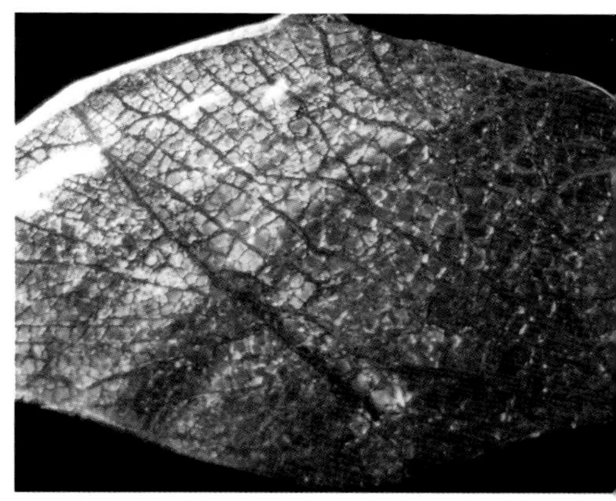

*Gregory Baker – Provincial Museum of Alberta*

*Fragment of ammonite shell reveals the beautiful iridescent colors of the ammolite gemstone.*

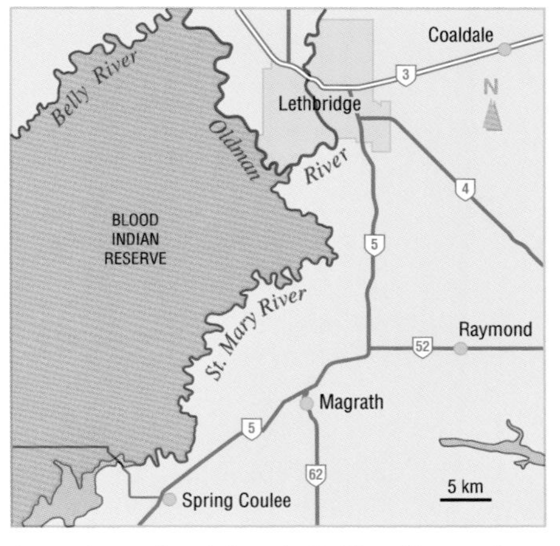


*Ammonites are found along the St. Mary River, southwest of Lethbridge.*

# GLOSSARY

**alkaline lake** A *salt lake* whose waters contain large amounts of dissolved sodium sulphate, sodium carbonate, sodium chloride, and other salts which collectively shift the pH to the alkaline side of neutrality.

**alluvial fan** A low, fan-shaped deposit of loose **sediment** built where a stream undergoes an abrupt reduction in slope.

**ammonite** An extinct marine shellfish whose living relatives include squids, cuttlefish,octopus, and the coiled Nautilus.

**anticline** An arch-shaped folding of the **strata** where the **limbs** are bent downward from the crest. The oldest rocks are in the centre of the fold.

**aragonite** A carbonate mineral ($CaCO_3$) occurring in **sedimentary rocks** and in shells. It is similar to **calcite** but has a different crystal structure.

**basement** The **igneous** and **metamorphic bedrock**, usually of **Precambrian** age, that lies beneath the younger **sedimentary rocks** in a region, i.e., where the Canadian Shield is in subsurface. The basement forms the core of all continental plates.

**bed** The smallest individual unit, or layer, of **sedimentary rock**. Each bed is marked by well -defined planes from the beds above and below it.

**bedding plane** A planar surface separating two layers, or **beds**, of **sedimentary rock**.

**bedrock** A general term for the rock, usually solid, that underlies the soil or other **unconsolidated** deposits that are on or near the surface of the earth.

**bentonite** A soft, plastic, light-colored **clay** formed from the decomposition of volcanic ash and largely composed of the clay mineral montmorillonite.

**bitumen** A mixture of semi-solid **hydrocarbons**; tar is an example.

**blowout dune** A sand dune with a long, scoop-shaped form, convex in the downwind direction and horns pointing upwind. Usually covered with sparse vegetation and found in areas of strong winds and abundant sand. Blowout dunes have formed since the last Ice Age and are particulary well developed in Western Canada.

**brachiopods** Marine invertebrates in which the soft body is enclosed by two unequal shell halves. They usually live attached to the sea bottom by a stalk. Most types are extinct.

**breccia** Similar to **conglomerate** except the rock fragments tend to be angular; may be **sedimentary** or formed by crushing or grinding along **faults**. There are also breccias of volcanic origin.

**brine** Water containing a high concentration of dissolved salts, especially common salt (halite).

**butte** A steep-sided, flat-topped hill, usually isolated, that is bordered by **talus** and formed by **erosion** of flat-lying **strata** and capped with a resistant layer of rock that protects the underlying softer rocks from erosion.

**calcareous** Rocks that contain or consist of as much as 50 per cent calcium carbonate ($CaCO_3$).

**calcite** The mineral composed of calcium carbonate ($CaCO_3$) that forms **limestone**. It is similar to **aragonite** but has a different crystal structure.

**carbonates** A general term for **sedimentary rocks**, such as **limestone** and **dolostone**, that are formed by **precipitation** of mineral carbonates.

**chalcedony** A **cryptocrystalline** variety of quartz; occurs as crusts with a rounded, mammillary surface and may be translucent or semitransparent. It has a wax-like lustre, and can be white, pale-blue, brown, black, or grey.

**chert** A hard, dense rock composed of extremely tiny, interlocking quartz crystals. It can be found in a variety of colors.

**clay** Clay is a rock particle size that is less than 1/256 millimetre in diameter. Clay is also a term used for a group of sheet-like minerals.

**conchiolin** A fibrous protein that makes up the organic basis of most **mollusc** shells.

**concretion** A hard nodule formed by the accumulation of mineral matter around an usually organic nucleus, such as a shell fragment.

**conglomerate** A coarse-grained **sedimentary rock** composed of rounded rock fragments larger than two millimetres in diameter set in a fine-grained cemented **matrix** of **sand** or **silt**. The consolidated equivalent of gravel.

**cross-bedding** Beds in sandstones that are inclined to each other and to the main sandstone layer. Cross-bedding in sand is formed by the action of either wind or water.

**cryptocrystalline** The texture of a rock or mineral consisting of crystals that are too small to be seen even under a microscope.

**crystal** A solid material in which the natural flat surfaces result from a regular internal arrangement of atoms. Where there is enough room to grow, most minerals can form crystals.

**crust** The thin outermost layer of the earth. It varies in thickness from about five kilometres beneath the oceans to about 60 kilometres beneath mountain chains.

**dip** The angle between a horizontal plane and a surface such as a **bedding plane**, **fault**, or **joint**.

**divide** A ridge separating two adjacent drainage basins.

**dolostone** A **sedimentary rock** composed mainly of the mineral **dolomite**.

**dolomite** An often creamy-white **carbonate** mineral composed mainly of magnesium and calcium.

**drumlin** A smooth, streamlined hill formed beneath a flowing **glacier** and usually composed of **till**. Drumlins are elongate in the direction of glacial flow.

**dune** A mound, hill, or ridge of wind-blown **sand**, either bare or variably covered by vegetation.

**dyke** A sheet of **igneous intrusive rock** that cuts across older rocks either vertically or at an angle. Also spelled dike.

**erosion** The process in which rocks

and **sediments** are broken down, loosened, and moved from one place to another by wind, water, ice, or gravity.

**erratic** A boulder or other large rock fragment transported by glacial ice to an area removed from that rock's source.

**escarpment** A steep slope or cliff forming a linear feature.

**evaporites** Sedimentary rocks formed by **precipitation** from evaporating sea water or salt lakes. Evaporites are classified according to chemical composition, e.g., gypsum, halite, and a variety of potash salts.

**extrusive/volcanic (igneous)** Magma that has hardened on the Earth's surface.

**fault** A fracture or a zone of fractures in the Earth's crust with relative movement between the rock masses on both sides.

**foliation** When layers separate or split into thin layers because of the segregation of different minerals into layers. Found in **metamorphic rocks** such as **schist**.

**Foothills** The region of low, rounded hills that are bounded on the east by the Interior Plains and on the west by the Front Ranges.

**formation** The basic unit in geology for the naming of an assemblage of rocks. Each formation represents a similar depositional environment; is large enough to be mapped at the surface; and is traceable in the subsurface, e.g., Cardium Formation.

**Front Ranges** The northwesterly trending mountain ranges of the Rocky Mountains that are bounded by the **Main Ranges** to the west and the **Foothills** to the east. The rocks are primarily **limestones**, **dolostones**, and **shales** of Paleozoic age.

**glacier** A large mass of ice, typically showing movement away from its source and formed from the compaction of snow.

**group** A group consists of two or more **formations**, e.g., Edmonton Group.

**gypsum** A widely distributed sulphate mineral formed by the evaporation of seawater by the action of ground water on sulphide-rich shales or by the alteration of anhydrite.

**halite** An **evaporite** mineral composed of sodium chloride ($NaCl$). Also known as common salt, rock salt.

**hanging valley** A former glacial valley which enters another larger glacial valley at a higher altitude.

**hydrocarbons** Organic chemicals composed of hydrogen and carbon atoms arranged in rings or chains.

**igneous rock** A rock that has solidified from **magma**. Magma that has hardened on the Earth's surface is called **extrusive (volcanic) rock**; magma that has hardened beneath the Earth's surface is called **intrusive (plutonic) rock**.

**intrusive/plutonic (igneous)** Magma that has intruded into pre-existing rocks beneath the Earth's surface and hardened.

**iridescent** Colors that are produced by the interaction of light and the layers of some minerals. As the angle of the light entering the mineral changes, a variety of different colors are produced. Examples include ammolite, opal, and mother-of-pearl.

**joint** A large fracture plane in a solid piece of rock in which there is no relative movement of the two sides, unlike a **fault**.

**karst** A type of topography that is formed mainly on **limestone** and **gypsum** by the dissolving action of water and is characterized by sinkholes, caves, and underground drainage.

**lava** Magma that flows from a volcano or fissures in the Earth's surface. Generally, the less viscous the lava the faster the flow, and the more viscous the lava the greater the tendency towards an explosive eruption.

**limb** One side of a **syncline** or **anticline**.

**limestone** A **sedimentary rock** composed of greater that 95 per cent calcium carbonate, mainly in the form of the mineral **calcite**.

**magma** Molten rock that forms **igneous rocks** upon cooling. Magma that reaches the surface is called lava and forms **volcanic (extrusive) rocks**; magma cooled at depth forms **plutonic (intrusive) rocks**.

**Main Ranges** In Alberta, the northwesterly trending mountain ranges of the Rocky Mountains that extend from the continental divide on the west to the **Front Ranges** on the east. The rocks are primarily **limestones**, **dolostones**, and **quartzites**.

**mantle** The layer between the Earth's **crust** and core. It is approximately 2300 kilometres thick.

**mass wasting** The downslope movement of soil and rock predominantly under the force of gravity, for example landslides. (synonymous with mass movement)

**massive rock** A rock that has a homogeneous appearance and is only occasionally broken by cracks, bedding, **foliation**, or **joints**.

**matrix** The relatively finer-grained material of a rock in which coarser particles are embedded.

**meander** A bend or curve in the course of a river which continuously swings from side to side in wide loops as it flows across flat country.

**metamorphic rock** A rock formed, without melting, from a pre-existing rock in response to pronounced changes of temperature, pressure, and chemical environments.

**metamorphism** The process by which rocks are altered in composition, texture, or mineral content without melting in response to changes in heat, pressure, and the introduction of new chemical substances.

**mica** A group of minerals that have sheetlike structures, low hardness, and readily split into thin, tough plates with a pearly lustre.

**molluscs** A large group of invertebrates that have unsegmented soft bodies and are usually covered with a hard shell. Examples are clams, snails, mussels, and oysters.

**moraine** The debris or rock fragments that were deposited by glaciers. The debris differs from water-laid deposits in that it is unstratified and the rock fragments are highly variable in size. There are many names attached to moraines depending on where and how they were formed.

**oil field** An underground accumulation of oil, and sometimes gas, that is prevented from escaping by an impermeable rock that forms an oil trap. An oil field is made up of two or more oil pools and is of economic value.

**oil trap** A rock structure or formation in which **hydrocarbons** accumulate and are prevented from escaping.

Also known as a trap.

**ore** A mineral occurring in sufficient quantity and quality to permit its economic extraction.

**outcrop** An exposure of **bedrock** at the Earth's surface.

**paleontology** The science that deals with fossil remains of both animals and plants. The information gained from this can be used to interpret ancient environments.

**percolating** The action of groundwater that seeps, oozes, or filters through soil without a definite channel.

**permeability** A measure of the ability of a porous rock, soil, or sediment to transmit fluid through pores, cracks, and **bedding planes**.

**petroleum** A naturally occurring oily, dark-colored **hydrocarbon** that is usually liquid but can be gaseous, solid, or a combination of states.

**pinnacle** A high tower or spire-shaped peak of rock, either isolated or at the summit of a hill or mountain.

**plateau** A flat, extensive upland area bounded by mountain ranges or steep slopes.

**point bar** The crescent-shaped accumulation of sand and gravel deposited on the inside of a **meander** as the stream velocity decreases.

**pore space** The spaces in a rock that are not occupied by solid material; includes fractures, voids, vesicles, and spaces between grains.

**porosity** The percentage of a rock or **sediment** that is pore space. Porous rocks are not necessarily permeable.

**potash** A general term that includes a variety of highly soluble potassium salts that precipitate from sea water when about 98 per cent of its volume has been evaporated.

**Precambrian Eon** The segment of time that began with the consolidation of the Earth's crust and ended about 544 million years later with the beginning of the Cambrian Period. This represents 88 per cent of the total length of geological time.

**precipitation** The process by which a substance is separated out from a solution as a solid.

**quartzite** A very hard, often white rock consisting almost completely of silicon dioxide and usually formed from the **metamorphism** of pure quartz sandstone.

**reserves (petroleum)** Identifiable resources of crude oil or natural gas in a subsurface rock which can be extracted profitably with present technology and under existing economic conditions.

**reservoir rock (petroleum)** Any subsurface rock that is porous and permeable and yields oil or gas. The most common reservoir rocks are **limestone**, **dolomite**, and **sandstone**.

**rock flour** An extremely fine glacial **sediment** formed by abrasion of rocks at the base of a **glacier**.

**sand** A common rock particle between 2 and 1/16 millimetre in diameter.

**sand and gravel bar** An accumulation of sand and gravel along the sides of a river or within the channel.

**schist** A type of **mica**-rich **metamorphic rock** that has strong **foliation**, allowing it to be easily split into thin slabs.

**schistosity** Where rocks split along parallel planes because of the distribution and arrangement of platy **mica** minerals. Found in **metamorphic rocks**, such as **schist**.

**sediment** Material derived from pre-existing rock and deposited at, or near, the Earth's surface.

**sedimentary rock** A consolidated rock formed of cemented **sediments**, usually in layers.

**shield** Where the basement rocks are exposed over large areas, such as the Canadian Shield. Also known as continental shield.

**shale** A fine-grained **sedimentary rock** formed by the cementation of **clay** and **silt** (mud).

**silt** Rock particles between 1/16 and 1/256 millimetre in diameter.

**sinkhole** A small, closed, subsurface depression formed in areas of **karst** topography and produced by dissolving **limestone** and **gypsum** or by the collapse of cave roofs. Surface water often flows down into the sinkhole and into an underground drainage system.

**slate** A dense, fine-grained **metamorphic rock** formed from a **shale**. Slate splits into thin, smooth plates and is often used as roofing tiles.

**spring** A place where underground water flows from a rock or soil onto the land or into a body of surface water.

**stratified** Stratified rocks are those in which the original bedding or layering formed at the time of deposition can be recognized.

**stratum** (plural **strata**) An individual bed or layer of **sedimentary** material, either rock or soil, that is visibly separable from the layers above and below it. There is usually a series of parallel layers, called strata. (synonym: bed)

**striations** A group of parallel or subparallel grooves or lines that were carved on bedrock and boulders by rocks embedded at the base of a flowing **glacier**. Striations indicate direction of ice movement.

**strip mining** A method of surface mining in which the ground cover is removed to obtain the economic material.

**stromatolite** A concentric laminated structure of calcium carbonate that was formed as a result of the growth of cyanobacteria, particularly during the **Precambrian Eon**. Stromatolites have a variety of forms, such as horizontal, columnar, domal and subspherical.

**stromatoporoid** An extinct colonial organism likely belonging to the phylum Porifera (sponges). Stromatoporoids built massive, laminated, calcareous structures, particularly during the Silurian and Devonian periods, that are important petroleum reservoirs in Alberta today.

**sulphate minerals** Minerals that contain the sulphate ions which are composed of one sulphur and four oxygen atoms.

**syncline** A U-shaped or V-shaped fold in the **strata** in which the **limbs** are concave upward and the oldest rocks are in the centre.

**talus** Rock fragments of any size derived from and accumulated at the base of a cliff or steep slope.

**thrust fault** A **fault** with a **dip** of less that 45° in which the block of rock above the fault plane moves up and over the lower block, so that the older rocks sit on top of the younger rocks. Thrust faults are characterized by horizontal compression.

**thrust sheet** The thick sheet of **strata** above a **thrust fault**. In the

Canadian Rockies, thrust sheets have been shoved eastward and over top of younger rocks.

**till** An unsorted, unstratified, and usually **unconsolidated sediment** deposited by a glacier, and containing all sizes of rock fragments from boulders to **clay**.

**U-shaped valley** A valley with steep upper walls that grade into a flat, wide valley floor,usually formed by glacial erosion.

**unconformity** A gap in the geologic record between two adjacent rock **formations**, indicating a period of **erosion** or of nondeposition; a break in the stratigraphic sequence.

**unconsolidated** A **sediment** that is loosely arranged, that has particles that are not cemented together.

**vein** An infilling in a fracture in a rock by one or more minerals. Veins can be of economic value, e.g., gold-rich quartz veins.

**vug** A small cavity in a rock that is usually lined with mineral crystals that may or may not be different in composition from the enclosing rock.

**weathering** The disintegration of rocks and minerals on the Earth's surface by either physical processes (such as rain, wind, heat, frost) or by chemical and biological processes. There is little or no transportation of the altered and loosened rocks.

# FURTHER READING

Beaty, C. *The Landscapes of Southern Alberta: A Regional Geomorphology.* Lethbridge: University of Lethbridge Printing Services, 1984.

Flygare, H. *Sir Alexander Mackenzie Historic Waterways in Alberta.* Canada: Banff Crag and Canyon, 1983.

Gadd, B. *Handbook of the Canadian Rockies*, Second Edition. Jasper, Corax Press, 1995.

Godfrey, J.D., ed. *Edmonton Beneath Our Feet: A Guide to the Geology of the Edmonton Region*. Edmonton: Edmonton Geological Society, 1993.

Hardy, W.G. (Editor-in-Chief), *Alberta, A Natural History*. Edmonton, Alberta: M.G. Hurtig Publishers, 1967.

Harrison, John E. *Evolution of a Landscape: The Quaternary Period in Waterton Lakes National Park.* Geological Survey of Canada Miscellaneous Report 26, 1976.

Horner, J.R.. and Gorman, J. *Digging Dinosaurs*. New York: Harper Collins Publishers, Inc., 1988.

Jackson, L.E., and M.C. Wilson, eds. *Geology of the Calgary Area.* Calgary: Canadian Society of Petroleum Geologists, 1987.

MacDonald, Janice, E. *Canoeing Alberta*. Edmonton: Lone Pine Publishing, 1985.

Mossop, G., ed. *Geological Atlas of the Western Canadian Sedimentary Basin*. Calgary: Canadian Society of Petroleum Geologists, 1994.

Nelson, S.J. *The Face of Time*. Calgary: Alberta Society of Petroleum Geologists, 1970.

Nelson, S.J. *The Geological History of the Interior Plains*. Prairie Forum, Vol. 9, No. 1984.

Royal Tyrrell Museum of Palaeontology, *The Land Before Us; The Making of Ancient Alberta*. Red Deer: Red Deer College Press, 1994.

Russell, D.A. *A Vanished World.* Ottawa: National Museum of Natural Sciences; and Edmonton: Hurtig Publishers, 1977.

Russell, D.A. *An Odyssey in Time: The Dinosaurs of North America.* Toronto: University of Toronto Press, 1989.

Yorath, C.J. *Where Terranes Collide.* Vancouver: Orca Book Publishers, 1990.

Yorath, C., and Gadd, B. *Of Rocks, Mountains and Jasper: A Visitor's Guide to the Geology of Jasper National Park*. Toronto: Dundwin Press, 1995.

# INDEX TO GEOLOGICAL SITES AND SELECTED GEOLOGICAL TOPICS

# REPRESENTATION, REALISM & FANTASY

**Key concepts for analysing film and television**

Jackie Newman and Roy Stafford

Published by *bfi* Education in association with *in the picture*

**Jackie Newman** is currently Head of Creative Arts at Ryburn Valley High School, Halifax, and **Roy Stafford** is a freelance media lecturer in West Yorkshire. The authors have worked together on Media INSET courses for English teachers and on film studies courses for adults. They have both worked as examiners in Film and Media Studies and written a wide range of media education teaching and learning materials.

*in the picture* publishes a range of material for teachers and students, including a magazine for media teachers. www.itpmag.demon.co.uk

Editor: Wendy Earle

Design: Alex Cameron

Film stills: Courtesy of *bfi* Stills
*Buffy the Vampire Slayer* (TV Series)
*The Crucible* (Twentieth Century Fox Film Corporation)
*Kes* (Woodfall Film Productions)
*To Kill a Mockingbird* (Pakula-Mulligan Productions)
*Pleasantville* (New Line Cinema)
*The Wizard of Oz* (Leow's Incorporated)

Printed in Great Britain by Cromwell Press



First published in 2002 by
*bfi* Education, 21 Stephen Street, London W1T 1LN

The British Film Institute offers everyone opportunities to experience, enjoy and discover more about the world of film, television and moving image culture.

www.bfi.org.uk/education

British Library Catalogue in Publication Data
A catalogue record for this book is available from the British Library.

ISBN: 1–903786–07–X